The Unofficial Trial of

Alexandra Morton

Other Books by Scott Renyard

Illustrated Screenplays

Who Killed Miracle? (2022)

The Pristine Coast (2023)

The Herring People (2023)

Trial of an Iconic Species (forthcoming 2023)

Children's Books

The Flag That Flew Up (2021)

The UNOFFICIAL TRIAL of ALEXANDRA MORTON

an illustrated screenplay

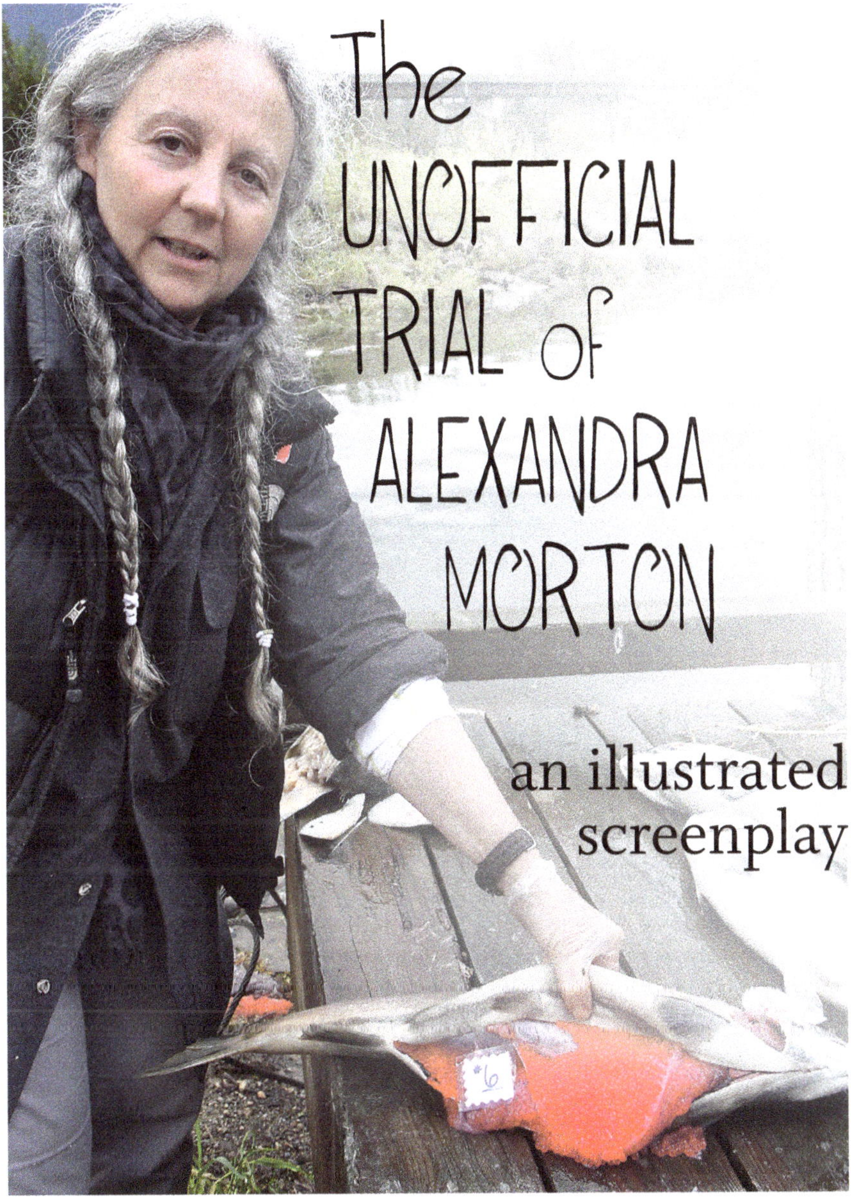

SCOTT RENYARD

juggernaut CLASSICS

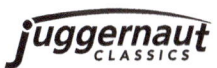

Published by Juggernaut Classics Inc.
Contact: scott@juggernautpictures.ca

ISBN: 978-1-998836-52-9 (softcover)
ISBN: 978-1-998836-53-6 (eBook)

Cover photograph by Scott Renyard.

Edited by Lesley Cameron

Cover design by Rob Neilson and Jan Westendorp
Book design by Jan Westendorp/katodesignandphoto.com

Lyrics for "Feedlot Blues" reproduced courtesy of Holly Arntzen and Kevin Wright

Juggernaut Classics Inc.

Contents

Introduction

In the spring of 2010, I spent several weeks following the Get Out Migration protest down Vancouver Island for my film *The Pristine Coast*. I then filmed portions of a canoe paddle protest down the Fraser River between October 20 and October 25 that same year. The paddle protest was timed to arrive in Vancouver at the Burrard Civic Marina at the entrance to False Creek on the first day of the Commission of Inquiry into the Decline of Sockeye Salmon in the Fraser River, also known as the Cohen Inquiry. The paddle protest participants joined other protesters, and together they marched across the Burrard Street Bridge to the lawn of the Vancouver Art Gallery at the corner of Georgia and Hornby Streets. The art gallery has a large open space that can accommodate a crowd and has long been the go-to venue for protests in Vancouver—and it was just a block away from where the Cohen Inquiry was taking place.

Alexandra Morton was a key participant in both protests—and an important character in *The Pristine Coast*—so I was there with my camera to cover as much of the event as possible in the hope that I would find a great ending for my film. The speeches were powerful and emotional, but the inclement weather made it very difficult to get the shot I was looking for. After recording Alexandra's speech in the pouring rain, I decided to pack up the camera and cross the street to the courthouse and sit in on the afternoon session of the Inquiry. That decision had a major impact on the next few years of my life.

I happened to sit in the back row of the courtroom, next to the director of Communications for the Inquiry, Carla Shore. On the first break, I asked her if I could come into the courtroom and film it while it was empty. I thought it might make an interesting closing shot for *The Pristine Coast*. Shore asked me to meet her at her office after the day's session had ended. At that meeting, I told her I was working on a film about wild salmon, and she said that I was welcome to use the designated camera position *during* the hearing. She told me to show up the next day at 7:30 AM so I could set up before the day's hearing started.

The camera position turned out to be right next to the Commissioner at the front of the courtroom. It was intimidating to say the least. I had never been in a courtroom in my life (unless you count traffic court, which I don't) and now I was suddenly right in the middle of the action. I wasn't just a fly on the wall; I was a fly in the middle of the dinner table. As I was setting up, Shore said, "Oh, by the way, you can't leave your position except on the breaks, even for the washroom." Yikes! So managing fluid intake also became part of my routine.

The terms of reference allowed only one camera in the courtroom at a time, but a line feed from the "in courtroom camera" position was available to send the recording live to a room outside the courtroom. This setup basically allowed multiple news outlets to get what they needed to report on the Inquiry without creating a potentially disruptive media scrum in the hearing room. For the majority of the time, however, I was the only media presence at the hearings. Most of the science and stakeholder concerns, which were the focus of the bulk of the hearing days, did not draw a lot of attention. As I packed up at the end of my first day, I asked if I could come back the next day. Since no other outlet had submitted a filming request, the answer was yes. After the third day, I asked, "How often can I come?" The response? As often as I liked. I was thrilled not only to be given the opportunity to film the testimony live, but also to be able to record footage that would have a consistent style to it.

Another key restriction was that I could only film while the Inquiry was in session. In other words, the rules required that I only turn on the camera when the Commissioner entered the room at the beginning of the day and then turn it off during breaks and as soon as the hearing was officially adjourned for the day. This was to prevent the accidental recording of private conversations between counsel and the stakeholders outside of the official, public hearing periods. That made sense, and I was careful to comply.

I immediately decided to reorganize my life and my work schedule and committed to filming as many of the hearing days as I could. After a few days, the Inquiry staff accepted me as one of their own. I was able to adjust the lighting and make other minor changes in the room so I could get an unobstructed view of the participants. I was allowed to adjust the blinds on the windows during breaks to control how much light came in, for example. It certainly wasn't like having a full lighting crew on a film set, but the adjustments made a huge difference to the quality of the footage.

Ideally, the hearings would have been a five days per week affair. However, not all weeks were full. Sometimes participants were not available or the legal teams for various stakeholders needed more time to prepare. In the end, 133 days of testimony took over a year to complete, and I was able to record 119 of them. But even with days off and breaks, my schedule was gruelling. A typical day of filming would start for me at 6:00 AM. I'd pack up my camera and gear, travel downtown, and be setting up by 7:30 AM. The testimony would usually finish around 4:00 PM, sometimes a little earlier. After a day on my feet, I would race home to dump the footage onto my hard drives. My practice is to always create three copies of any day's footage. Hard drives are notorious for failing, and anything less than three copies would be risky. The first backup of the Inquiry footage would take about four hours and finish by 8:00 PM. I would then move straight onto the second one so that it would be finished around midnight. Before I went to bed, I would set the computer in motion to make the third copy overnight. In the morning, I would check that the footage was backed up and then reformat my camera cards for that day's work.

The limit of one camera in the hearing room at any one time created an added challenge for me. In this situation, it's always best to capture all of the question and all of the answer to it on camera. In other words, for the best coverage of the event, I wanted the camera to be on the speaker from the moment they began to speak to the moment they stopped, and then on the next speaker to repeat the process. How did I do this with one camera? Speed. I had to frame and focus with the same motion. I remember seeing skilled cameramen do this type of manoeuvre during my days on Hollywood shoots. I never thought I would have to try to replicate this skill at a live event. But here I was, bumbling through the first few days until I got better at it. And in this case, I only had one chance to capture the footage, all while being as quick and discreet as possible.

Protocol determined the order in which people questioned the members of the panels—the Commission counsel went first, followed by the provincial and federal governments, and then other stakeholders—which was helpful, but anyone could interrupt with an objection or added comment. I was literally on my toes for the entire day, and as the days passed, I guess my camera moves became better. At least, that's what my editors told me. The Commission counsel, who spoke first, was located close to the members of the panel in the room. As a result, my camera swings were short and relatively easy to land on the speaker in focus before they began to talk. Crown counsel, which included the provincial and federal governments' counsel, were located in the middle of the room with the BC Salmon Farmers Association counsel. For interactions between these lawyers and the panellists, the camera swings were larger but still quite manageable. But for interactions with some of the other stakeholders, and particularly the counsels for the NGOs, I had to resort to rapid whip pans back and forth across the room.

Luckily, most speakers took a moment to gather their thoughts before asking or answering a question. And usually, that brief pause was all I needed to record the subject's entire dialogue on camera.

In the early days of the Inquiry, I wasn't sure the footage would result in an interesting ending for *The Pristine Coast* or if I would be able to use it in some way in another film. But it's not every day an independent film maker gets a chance to record an inquiry, so I decided I would continue to record it for mostly altruistic reasons. I knew that many of the fights between stakeholders over wild salmon had been going on for decades. I hoped that my footage would, in some way, give stakeholders another tool that might help them reach a consensus not only on how to save BC's wild salmon populations, and sockeye in particular, but also on how to return them to their once abundant levels. I had no funding for this project, so I basically volunteered my time and spent about $20,000 of my own savings to capture and store all the footage.

While most of the Cohen Inquiry didn't make the headlines, two participants attracted a lot of attention: Alexandra Morton and Dr. Kristi Miller. Both women had been in the spotlight before. Morton, an independent biologist, had been researching and raising awareness about the impacts of fish farms on wild salmon near her home in the Broughton Archipelago since the 1980s. She essentially exposed the industry's connection to the decline in wild salmon populations. And Miller, an expert in genomics from the Department of Fisheries and Oceans, was the lead author of an article published in *Science* in January 2011 titled "Genomic Signatures Predict Migration and Spawning Failure in Wild Canadian Salmon." The article represented a breakthrough in an understanding of the underlying cause of wild salmon declines, and its publication was swiftly followed by the Canadian Prime Minister's Office effectively muzzling Miller by barring her from speaking to the media.

I began to scope out the days these two key participants would appear at the Inquiry. Both were scheduled for two days each. Miller would appear on August 24 and 25, 2011, and Morton, September 7 and 8, 2011. As the Miller and Morton appearance dates grew closer, I was concerned that I could be bumped from "my" camera position by a larger broadcaster like the CBC. They had set up all of their audio and video equipment in advance, and I thought that meant they were planning on filming some of the Inquiry at some point. I was fairly sure they would be most interested in the Miller and Morton days, so I decided to ask if I could be the feed camera to the media room. The staff agreed to my request once they determined my camera could handle the feed to the media room. By this time, I was beginning to think there was a film in all the footage I had been gathering.

The look and feel of the footage that I had gathered for most of the year could have been lost if other camera operators suddenly took the lead. On the few days I was bumped, it was clear that a news channel had very different needs for its stories than I did for mine. It appeared to me that news channels were only looking for a few short clips to cover the narrative of a brief news story. To do that, I noticed they would record for a few minutes, then turn off the camera and wait for the next break to pack up and leave. If I had been bumped for more than the occasional day, it would have had two significant consequences for me: I would have lost the tighter framing of the subjects I preferred, even if I could have acquired the wider-framed news footage through the feed, and I might have had big gaps in the testimony at key moments if the in-court camera was off. And had I been bumped for the critical Morton and Miller days, there is no question that the narrative of any long-form film I decided to make with the footage would have been significantly compromised. In fact, I likely would have had to shelve any plans for a long-form film. Thankfully, though, I was able to remain in "my" spot, and I recorded all of the testimony on the crucial days.

Every day the Inquiry sound technician supplied a direct audio line feed to my camera. All of the testimony for the Inquiry was being audio-recorded for two reasons: the transcribers who type the live testimony during a hearing need to check their work before submitting the final transcript, and the testimony was being translated into French, Canada's other official language. The line feed to my camera meant I could record sound from all of the individual microphones used by the participants around the room. This provided nearly perfect sound for all the testimony. For 115 days of filming, the line feed to my camera was perfectly clear. However, on the crucial four days, the signal was nothing but static and unusable. For some reason, a different sound technician was on duty during the days when Dr. Kristi Miller and Alexandra Morton spoke. The loss of my audio line feed was very odd, and when I asked for the line to be checked, a solution could not be found. At that time, it was no secret that the Harper government had already muzzled scientists like Miller, whose research was suddenly the target of message control. It was difficult not to feel suspicious about the timing of the change in technician and the sudden loss of the audio feed. Fortunately, I was working on the assumption that a glitch in the line feed could occur at any time and was recording sound on another track with a narrow cone boom microphone. The sound was almost as good as the line feed, and I was able to capture the Miller and Morton testimony effectively.

It wasn't until the hearings were over that I realized Alexandra Morton's testimony was actually a film in its own right. Initially I thought it would just be part of a narrative that would include the rest of the witness testimony for my film on the Cohen Inquiry, *Trial of an Iconic Species*. But it was clear from the cross-examinations that the Province of British Columbia, the Government of Canada, the BC Salmon Farmers Association, and the Aboriginal Aquaculture Association saw Morton as their main adversary at the Inquiry. She was the biggest threat to the industry, and her years of raising awareness were likely a main driver for the Cohen Inquiry in the first place. Apparently the public thought so too. During most of the hearing days, the gallery was empty. But when Morton was on the panel, the courtroom gallery was packed. The exchanges were compelling and tense. The Cohen Inquiry was somewhat unusual in that it was like a trial. In the Supreme Court of Canada *Canada (Attorney General) v. Canada (Commission of Inquiry on the Blood System)* (1997) judgment, it states that "a commission of inquiry is not a court or tribunal and has no authority to determine legal liability; it does not necessarily follow the same laws of evidence or procedure that a court or tribunal would observe."[1] However, Brock Martland, one of the lawyers present, stated that the Cohen Inquiry was "trial-like" during one of the Morton days when the spectators in the gallery began to boo and get restless because they didn't like the questions posed to Morton. It really was like a Hollywood courtroom drama, complete with outbursts from the gallery that compelled the judge, or in this case the Commissioner and his counsel, to demand, in a polite Canadian way, order in the court.

After a couple of weeks of filming, I began to realize there were distinct alliances between some of the stakeholders. Although one of the terms of reference and guiding principles for the Commissioner was to foster broad cooperation between the stakeholders, the underlying tension between the groups for and against the aquaculture industry was very evident. I'm not sure why it hadn't occurred to me previously, but I suppose I thought each stakeholder would be presenting arguments based solely on their own knowledge, research, or perspective. However, it was clear early on that the Province of British Columbia, the Government of Canada, the BC Salmon Farmers Association, and the Aboriginal Aquaculture Coalition supported each other's motions and objections and acted in concert. First Nations representatives formed another obvious-to-me alliance since their positions on most issues were similar and they often shared cross-examination time. NGOs were also aligned for obvious reasons. Additionally, it seemed to me that First Nations and environmental associations were for the

---

1   See https://scc-csc.lexum.com/scc-csc/scc-csc/en/item/1548/index.do

most part aligned—with one marked exception concerning conflicts between commercial and Indigenous fishing allocations and rights. During Morton's testimony, the alliance between the two senior levels of government and the aquaculture industry became even more obvious, and when Morton's lawyer, Greg McDade, was cross-examining the panellists, the legal representatives for these three groups rose up to object several times. Sometimes the objections involved rather long explanations and speeches, which seemed to be more about using up valuable hearing time (each speaker was given a set number of minutes on the stand) to reduce the number of points that McDade could make. Prior to this experience, I had always assumed that in this type of situation the government was a neutral party and responsible for mediating disputes between citizens and stakeholders. It had never occurred to me that hiring a neutral party to dispense with any bias or manipulation by government itself was a crucial feature of an inquiry. I suppose it wasn't a huge surprise that our senior levels of government and the interests of the industry were strongly aligned. They both have a huge stake in the money and the taxes the industry generates. But what was surprising was the open display of allegiance in the public forum which suggested to me that our governments were obviously less interested in the other stakeholders' concerns and the public's concern over aquaculture's impacts on their businesses and the environment.

The Cohen Inquiry was designed to investigate the reasons for the alarming decline of the Fraser River sockeye populations. It's important to note, though, that it was not tasked with finding fault or a "smoking gun," as it was referred to several times during the hearings. In the terms of reference, the "Commissioner [was] to perform his duties without expressing any conclusions or recommendations regarding the civil or criminal liability of any person or organization."[2] So in one way it didn't seem necessary for any stakeholder to try to block testimony

2   See https://www.canada.ca/en/news/archive/2009/11/terms-reference-commission-inquiry-into-decline-sockeye-salmon-fraser-river.html

from the Aquaculture Coalition (the coalition comprising Alexandra Morton and NGOs that were against open net pen fish farming), or anyone else for that matter. But clearly, even if the Inquiry wasn't to rule on liability, the stakes were very high for all concerned, and any information or documents that were included could sway government policy going forward. Each document presented had to be entered and accepted as an official exhibit before it could become part of the basket of documents that the Commissioner could use to write his final report. All stakeholders therefore had a powerful incentive to try to get documents important to them included as exhibits and, of course, to block the others from doing the same. It was fascinating to watch the gamesmanship that resulted from the rules of the Inquiry and how it played out in court. Apparently, half a million documents, most of them from the Department of Fisheries and Oceans, were submitted to the Inquiry as relevant. In the end, only 2,147 documents were made official exhibits. (If you read the final report, you'll see that it mentions 2,145 documents accepted as exhibits, but 2,147 were distributed during the Inquiry. I titled my library of Inquiry footage *Exhibit 2148* in honour of the 2,147 official exhibits. You can watch all 395 episodes of *Exhibit 2148* on The Green Channel.)

One objection from the government–aquaculture block occurred during Morton's testimony when a document she compiled was introduced by her lawyer, Greg McDade, for inclusion as an exhibit. The objection was that the information it contained was not Morton's opinion or research but a collection of information from other sources. Each stakeholder was given access to all the documents submitted to the Inquiry on a system called Ringtail. While she was waiting for her scheduled appearance, Morton decided to review as much of the material as possible and condense and compile what she felt was key evidence that the Commissioner should review. She was concerned that some of the information would not be included as an exhibit and believed that compiling it into one document would provide a summary of issues that ought to be considered by the Commissioner in writing his report. When there are objections to a document being included as an exhibit, it is classed as "marked for identification." This means it is set aside until the Commissioner can receive and evaluate the oral and written arguments for and against the document being allowed to stand as a full exhibit. In the end, the document Morton compiled was allowed to stand as a full exhibit as part of her perspective—but only after a lot of back and forth and the use of valuable hearing time that could have been used to introduce or discuss other evidence.

The Commissioner gathered information for the Cohen Inquiry in three main ways: 900 submissions from the public, some of which were as simple as an email; the commissioning of 15 technical reports; and the evidentiary hearings. In April 2010, five months prior to the beginning of the Cohen Inquiry, the Commissioner announced that 53 individuals, groups, and organizations that had applied individually would be clustered into 21 groups to present their testimony at the hearings. They were presumably grouped together because they had similar interests and clustering them in this way would reduce repetition and save valuable and expensive hearing time. But another interesting detail about participation emerged during the hearings themselves when the counsel for Canada stated that he didn't choose Morton to be on the Perspectives Panel. That comment suggested there was some attempt by stakeholders to challenge not only what panel participants could appear on, but also whether they should be allowed to appear at all. It also appeared that the Government of Canada preferred to keep an important player like Morton out of the Inquiry. In addition, each time a new panel was introduced, the Commission counsel spent a fair amount of time reviewing each participant's credentials and then recommending whether or not they should be accepted as witnesses qualified to speak on the topic. Most of the time the selection process was routine, but occasionally there would be an objection or question about a witness's credentials and whether they would be allowed to give testimony. In the end, 179 witnesses, who were all part of the 21 groups, were selected and subsequently gave evidence at the hearings.

Another interesting twist concerned Alexandra Morton. Morton had published many articles in peer-reviewed journals on various impacts of aquaculture on wild salmon and received an honorary doctorate from Simon Fraser University in 2010 for her work in this area. I was therefore very surprised to learn that she had not been invited to present her testimony as an expert witness on a science-based topic but was instead part of what was referred to as the Perspectives Panel. The Perspectives Panel included two speakers from environmental groups (Alexandra Morton is the director and founder of the Raincoast Research Society) and two from the aquaculture industry. This meant that Morton was supposed to give general comments as part of her environmental perspective, so she would not offer what would be considered an "expert opinion" on any of the science topics. Was Morton on the right panel? To this day, I still don't know for sure, but it certainly added drama for a film. The tension in the room as Morton and Catherine Stewart, executive director for CAAR, offered views that were diametrically opposite to those of the fish farm proponents, Mia Parker, former manager of Regulatory Affairs, Grieg Seafood, and Clare Backman, director of Compliance and Community Relations, Marine Harvest, was palpable.

I came away from the Cohen Inquiry with an appreciation for the legal teams and the work that must have gone into preparing for the hearings, the research, and, most of all, the challenges that must come with representing potentially unpopular positions. But the two days on which Morton spoke were not just about squabbles over facts or allocations of fish resources. The questioning became much more personal and reminded me of some recent election campaigns that resorted to personal attacks and tried to sidestep debate about substantive issues. I thought it was crucial that the public get to see how a woman who was essentially a whistleblower speaking out against very powerful men and organizations was treated during what, in Canada, is supposed to be an information-gathering affair. From where I was standing, it looked like the latter's strategy was based primarily on an effort to eat up limited testimony time and prevent some information from being considered by the Commissioner and on a thinly veiled attempt to intimidate Morton. I felt the intimidation effort sullied the prestige and importance associated with Canadian inquiries. I couldn't shake the feeling that if our governments and industry were resorting to these types of tactics, it was probably because there really isn't a good defence for the damage being done by open net pen fish farms.

Morton once told me that she was not just a scientist fighting for wild salmon but also a woman fighting for her home. There is no doubt in my mind that the aquaculture industry has had a profound negative impact not only on Morton's home of Echo Bay in the Broughton Archipelago region of the BC coast but also on the social, economic, and cultural fabric of many BC coastal communities. It is therefore not surprising that Morton has become a popular proxy for anyone who has experienced changes in their communities as a result of the loss of wild salmon. She has taken a stand against government and industry, often at a significant personal cost, to speak up for wild salmon. How many among us can say the same?

A Note from Alexandra Morton

Participating in the Cohen Inquiry into the Decline of the Fraser River Sockeye Salmon (the Cohen Inquiry) was like taking a step through a portal into the dark world of the bureaucracy in charge of regulating salmon farms in Canada. In 2010, when I was accepted as a registered participant of this inquiry, I was fully aware of the headwinds facing anyone who thought they could bring reason to the management of salmon farming. What I didn't realize was that Fisheries and Oceans Canada already knew that industrial salmon farming posed serious risks to wild salmon.

The Cohen Inquiry gave participants access to 500,000 government documents that were submitted by bureaucrats and government scientists and were considered to contain information potentially related to the 17-year decline of the Fraser River sockeye salmon. I quickly picked up the trail left by the DFO scientists who had looked at hundreds of thousands of superficially perfectly healthy sockeye salmon lying dead on the riverbanks just days before they should have spawned. Generally, when fish go missing they leave no trace, but now the bodies of beautiful ripe females full of eggs were strewn across Fraser River spawning grounds. The scientists called it *pre-spawn mortality*. Before the mid-1990s, pre-spawn mortality had been tightly associated with unusually warm river water. After the mid-1990s, huge numbers of salmon were apparently dying for random reasons throughout the watershed.

As I tunnelled through the massive document cache, the scientists' distress was unmistakable to me. They described some sockeye as "bleeders," gushing blood when touched. These DFO biologists worked through a list of possible causes, tossing out each theory as they looked closer at the evidence. Then Dr. Kristi Miller, who was head of the DFO genomics lab in Nanaimo and tasked with comparing the immune systems of the dying and healthy sockeye, reported that the millions of dying sockeye were fighting a virus linked to leukemia.

The only known salmon virus that met this description was salmon leukemia, which was first discovered by DFO scientists in the 1990s in the salmon farms in the Discovery Islands—directly in the path of Fraser sockeye. Miller noted that the one run of Fraser sockeye whose DNA is never found in the Discovery Islands, the Harrison sockeye, was affected. However, its population was not declining with the other runs; instead, it was rising. She also found the signature brain tumours reported with salmon leukemia in the dead salmon.

Then I read about what happened to her after she made these discoveries.

She was inexorably crushed until she had to abandon her work in this area. Her funding was not renewed. She arrived at the Cohen Inquiry flanked by large security guards via a back elevator so no one could talk to her. She was ridiculed and sidelined by DFO. I was incredulous. Over $20 million had been spent on the inquiry into what was killing Canada's most valuable wild salmon and the scientist who discovered the cause was silenced.

I carefully assembled Miller's work, the work of the biologists before her, and the systematic efforts to protect the salmon farming industry from culpability for the collapse of a biological, ecological, and economic engine that allowed the BC coast to flourish. My lawyers, Greg McDade and Lisa Glowacki, and I worked together to map the most efficient plan to communicate all these findings. We planned to submit my report as an exhibit and spend our limited time in front of Justice Bruce Cohen explaining the significance of Miller's findings. I arrived with a well-tabbed binder so I could refer to specific documents in a timely manner if Cohen had questions.

Government and industry had effectively blocked Miller. And they obviously had no intention of allowing me to breach their well-executed strategy: They leapt to oppose my report being entered as an exhibit. As I saw precious minutes ticking away, wasted by an orchestrated wall of opposition, I felt the pressure of tears of frustration build. I was embarrassed. Surely I would not cry on the stand! When we broke for lunch, I looped my arm in my son's and together we rushed outside. There I regained my composure and mentally pulled down a fencing mask. I would do my best to prevent the attacks from discrediting me and, by association, the report that held the well-protected response to a critical question: *What Is Happening to the Fraser Sockeye?*[3]

The salmon leukemia virus was most active in chinook salmon farms. With news seeping through the cracks that Miller had found evidence of this virus in millions of dying Fraser River sockeye salmon, the company farming chinook salmon in the Discovery Islands slowly replaced them with Atlantic salmon. While the Atlantic salmon were widely infected with a different virus, piscine orthoreovirus (PRV), the Fraser sockeye stopped dying of pre-spawn mortality.

Dr. Kristi Miller should have been awarded the Order of Canada. Instead, the story was buried. The salmon farming industry switched to Atlantic salmon and the Fraser sockeye went on to be infected with both PRV and tenacibaculum, a bacteria that causes mouth-rot in farm salmon. In December 2020, then Minister of Fisheries Bernadette Jordan bravely prohibited the restocking of all 19 salmon farms in the Discovery Islands. This is the most sensitive portion of the Fraser sockeye migration route because its narrow channels are easily saturated with pathogens from industrial salmon farming.

The summer of 2023 will see the return of the first generation of adult Fraser sockeye salmon that were not exposed to salmon farms

---

3   See https://tinyurl.com/mshct8n4

when they left the river to migrate to the open ocean as juveniles. However, in March 2023, the salmon farming industry and two First Nations sued the current Minister of Fisheries, Joyce Murray, over the decision to not renew licences for farms in the Discovery Islands. At time of writing, there has been no resolution to that legal action. Will this be the last generation of sockeye given a chance to thrive?

Today, I continue to read internal government communications using the *Access to Information Act.* The Cohen Inquiry taught me that if a government actively refuses to respond to evidence of a piece of our world being destroyed, there is a good chance a conversation is going on behind closed doors with an industry that has a vested interest in the topic. Unless you are privy to this conversation, you are swinging at windmills.

Thank you to Scott Renyard for recognizing the importance of the Cohen Inquiry into the Fraser River sockeye salmon and spending hundreds of hours capturing the collaboration of industry and government to suppress the truth.

Cast of Characters

(In Order of Appearance)

*At the time the documentary that this book is based on was filmed, Clifton Prowse, Mitchell Taylor, and Gregory McDade were all QCs. Since the death of Queen Elizabeth II in September 2022, that designation has changed to KC. I have opted to present them in this book as QC to maintain consistency with text that appears onscreen in the film version of *The Unofficial Trial of Alexandra Morton*.

ALAN BLAIR:
  Counsel, BC Salmon Farmers Association

CLIFTON PROWSE, QC*:
  Senior Counsel, Province of British Columbia

STEVEN KELLIHER:
  Counsel, Aboriginal Aquaculture Association

MITCHELL TAYLOR, QC*:
  Senior Counsel, Government of Canada

ALEXANDRA MORTON:
  Independent Biologist, Raincoast Research Society

HONORABLE BRUCE COHEN:
  Commissioner, The Uncertain Future of Fraser River Sockeye

LEONARD GILES:
Commission Registrar

BROCK MARTLAND:
Commission Staff Council, The Uncertain Future of Fraser River Sockeye

CATHERINE STEWART:
Former Campaign Director, Living Oceans Society

MIA PARKER:
Former Regulatory Manager, Grieg Seafood

CLARE BACKMAN:
Director of Environmental Compliance and Community Relations, Marine Harvest

GREGORY MCDADE, QC*:
Counsel, Aquaculture Coalition/Alexandra Morton

TIM LEADEM:
Counsel, Conservation Coalition

BRENDA GAERTNER:
Counsel, First Nations Coalition

The Unofficial Trial of Alexandra Morton
—an illustrated screenplay

By
Scott Renyard

Canada's Violations of the Fisheries Act and NAAEC
by Allowing British Columbia Salmon Feedlots
to Degrade Wild Salmon Habitat

Pink salmon fry from Broughton Archipelago parasitized by adult sea lice. Photo courtesy of Alexandra Morton

Citizen Petition
Submitted to the Commission for Environmental Cooperation
Under the North American Agreement On Environmental Cooperation

Submitted by:

**Center for Biological Diversity
Pacific Coast Wild Salmon Society
Kwikwasu'tinuxw Haxwa'mis First Nation
Pacific Coast Federation of Fishermen's Associations**

February 7, 2012

FADE IN:

JUGGERNAUT ANIMATED LOGO . . .

TITLE: Juggernaut Pictures presents . . .

TITLE: A Scott Renyard documentary

SUPERSCRIPT: Typing. "November 6, 2009. The Canadian Government commissions an inquiry into the rapid decline of Fraser River sockeye."

FADE TO:

INT. COURTROOM — DAY

Alan Blair, counsel for the BC Salmon Farmers Association, is at the microphone.

> MR. BLAIR
> Ms. Morton may choose to do so in the world of the blogs and the Web and endless postings, which we've, all could read if we chose to, um, but to make good that breach of code of ethics violation here, under oath, would be professional misconduct.

SUPERSCRIPT: Typing. "After 20 years of researching the impacts of fish farms on wild salmon, Alexandra Morton gets access to 500,000 government documents."

FADE TO:

MR. PROWSE
I don't think this is evidence. And, and asking what her "perspective," quote unquote, is I don't think really advances the matter. So I object to the line of questioning on that basis.

SUPERSCRIPT: Typing. "October 25, 2010. The first of 133 days of evidentiary hearings."

FADE TO:

MR. KELLIHER
You're the only one that isn't corrupted by business, by government, by a university. Is that correct?

SUPERSCRIPT: Typing. "September 7, 2011. Nearly a year after the hearings began, Alexandra Morton appeared as a witness for two days."

FADE TO:

                    MR. TAYLOR
You have no evidence to support
that accusation that people
in DFO do things just to keep
their job or don't tell the
truth just to keep their job,
do you?

PANNING OVER — TO MS. MORTON

                    MS. MORTON
I actually do, but I'm not
going to reveal all my sources,
because they're scared.

SONG BEGINS — FEEDLOT BLUES

PHOTO — Anti-fish farm protesters around
a Cohen Commission poster promoting
www.salmonaresacred.org.

                HOLLY ARNTZEN (V.O.)
Feedlot blues are taking my
home away.

PHOTO — Alexandra Morton making a speech with
Mr. Peter Julian, MP, looking on.

PHOTO — Morton standing with fellow
protesters. They are all in First Nations
regalia.

                HOLLY ARNTZEN (V.O.)
Feedlot blues tell me that I
can't stay.

A group of citizens supporting Alexandra Morton's appearance at the Cohen Inquiry pose around a sign encouraging more people to attend the hearings. (Photo credit: Anissa Reed)

Alexandra Morton takes a break from the Fraser River paddle protest and speaks to a crowd in New Westminster, BC, about the impacts open net pen farms are having on Fraser River wild salmon, September 16, 2011. Peter Julian, MP, and New Westminster city councillor Chuck Puchmayer attend the event. (Photo credit: John Preston)

Alexandra Morton, hereditary chief and master carver Beau Dick and his daughters, on their historic journey to break copper in Victoria, stand with a tray of farmed Atlantic salmon being sold at a supermarket to communicate the dangers of this product to wild salmon, February 7, 2013. (Photo credit: John Preston)

PHOTO — Morton on the shores of the Harrison River, taking samples from dead sockeye.

PHOTO — Morton, standing with her back to the camera, addressing a group of protesters.

> HOLLY ARNTZEN (V.O.)
> Wild salmon disappearing . . .

PHOTO — Morton on the steps of the BC Provincial legislature building.

> HOLLY ARNTZEN (V.O.)
> I've got to leave Echo Bay. Oh yeah . . .

FADE TO:

TITLE: The Unofficial Trial of Alexandra Morton

OVER BLACK

SUPERSCRIPT: Typing. "Examination in Chief begins."

INT. FEDERAL COURTROOM — DAY

The Commissioner, Bruce Cohen, and the registrar enter. Commissioner Cohen sits down.

LOWER THIRD: Hon. Bruce Cohen, Commissioner, The Uncertain Future of Fraser River Sockeye.

A protester holds up a sign at the finale of the Get Out Migration, which was led by Alexandra Morton to make politicians aware of how many people wanted salmon farms removed from the ocean. (Photo credit: Anissa Reed)

REGISTRAR GILES
The hearing is now resumed.

PANNING ACROSS — the courtroom past the panel
that includes, from left to right, Catherine
Stewart, Alexandra Morton, Clare Backman and
Mia Parker, to find Brock Martland, counsel
for the Commission.

COMMISSIONER COHEN
Mr. Martland?

MR. MARTLAND
Mr. Commissioner, we have today
a panel entitled Perspectives
on Management, Risks and
Finfish Aquaculture, and the
members of the panel from left
to right before you are, first,
Catherine Stewart from the
Living Oceans Society,

**LOWER THIRD: Brock Martland, Commission Staff
Council.**

MR. MARTLAND
. . . Alexandra Morton, the
Executive Director of the
Raincoast Research Society,
Clare Backman, the Director of
Environmental Compliance and
Community Relations with Marine
Harvest Canada, and Mia Parker
who, until recently, was the
Manager of Regulatory Affairs
with Grieg Seafood.

Alexandra Morton speaks to a group of concerned citizens at the Goldstream River just outside of Victoria, BC. Her talk was about the importance of wild salmon to the watersheds of rivers like this one and the damage done to the environmental balance by open net pen fish farms, September 17, 2011. (Photo credit: Anissa Reed)

Alexandra Morton takes samples on November 4, 2011, from a sockeye salmon that died before spawning and was drifting past in the Harrison River. (See more photos of the Harrison River die-off in Appendix B.) (Photo credit: Anissa Reed)

ON MS. MORTON

LOWER THIRD: Alexandra Morton, Independent
Biologist, Raincoast Research Society.

MR. MARTLAND (O.S.)
Ms. Morton, I'd like to start
with you.

DOCUMENT — Exhibit 1798, Alexandra Morton's
resumé. The front page pushes forward to
feature her name.

MR. MARTLAND (O.S.)
I'm going to first begin by
having your CV, which, Mr.
Lunn, is number 2 on Commission
counsel's list of documents for
this panel.

DOCUMENT — panning down to highlight "1978-
1979 Naval Oceans Systems Center sound
production of dolphins."

MR. MARTLAND (O.S.)
Since the late 1970s, you have
conducted research related to
marine mammals,

ON MS. MORTON

MR. MARTLAND (O.S.)
. . . including a long-term
field study of killer whale
ecology in the Broughton
Archipelago in this province.

ARTICLE — The front page of Morton, A.B.
2000. Occurrence, photo-identification
and prey of Pacific white-sided dolphins
(*Lagenorhynchus obliqiudens*) in the Broughton
Archipelago, Canada 1985-1997.

ARTICLE — Morton, A.B., and Symonds, H.K.
2002. Displacement of Orcinus orca by high
amplitude sound in British Columbia, Canada.

> MR. MARTLAND (O.S.)
> You published several academic
> articles relating to marine
> mammal ecology and behaviour.

ON MS. MORTON

> MR. MARTLAND (O.S.)
> And during the 1990s, you began
> to have concerns . . .

**HEADLINE — "Alexandra Morton: A watchful eye
on salmon farming," the *Fisherman News*.**

> MR. MARTLAND (O.S.)
> . . . about the pace of the
> development for . . .

PANNING UP — to reveal a photo of a young
Alexandra Morton.

> MR. MARTLAND (O.S.)
> . . . fish farms in the
> Broughton Archipelago,
> the . . .

• ALEXANDRA MORTON ... asking questions of government and expecting answers.

Alexandra Morton: A watchful eye on salmon farming

by Sean Griffin

Five years ago, when fishers like Gulf troller Bill Proctor and prawn fishers in the Broughton Archipelago first came to see whale researcher Alexandra Morton, she felt the first nagging worry that the salmon farms — introduced into the area only a couple of years earlier — were an environmental tragedy in the making. The prawns were disappearing from the surrounding area under the netpens and the numbers of spring and chum salmon seemed to be declining in the bays and inlets where salmon farms had been located.

Then last spring, when she saw the diseased spring salmon, again brought to her by local fishers, it seemed the confirmation of her worst fears.

"The fish all had enlarged spleens and gall bladders," she says. One fisher reported that of the 12 springs that he had caught, 11 were diseased. And the number of fish that returned to the waters of the Archipelago has been declining alarmingly, she adds.

We're talking in Victoria where Morton is taking a break from her whale research station — her floathouse in Echo Bay in Simoom Sound — to give a lecture at the museum and to promote the two children's books she has written, *Siwiti and In the Company of Whales.* Published a few years ago, their continued popularity is a testament to

Morton's painstaking research work and her enduring fascination with whales.

Today, however, the papers stuffed into the tote bag beside her at the table is testament to another cause she has taken up just as passionately.

The papers are copies of the letters she has written over the past five years to dozens of government agencies, ministers and deputy ministers — pointing to the environmental dangers that salmon farming seems to bring with it wherever it goes.

LETTERS TO DFO

Ironically, when prawn fishers initially came to her five years ago, Morton wrote the first letter to the Department of Fisheries and Oceans, hoping that it would

draw scientists' attention to the problem and something would be done — quickly.

Nothing happened. On the contrary, DFO biologists suggested that the problem of declining prawn stocks around salmon farms was probably "just a quirk of prawn biology," Morton recalls.

Not prepared to be put off, she began doing her own research, and continued writing letters. By 1993, with a growing body of scientific material underscoring her doubts about salmon farming, her letter-writing had become a campaign.

The letters have appeared in various publications, including The Fisherman, but mostly they're letters directed at Victoria and Ottawa, prodding government officials and asking pointed questions.

14 • THE FISHERMAN / JULY 25, 1994

An article written by Sean Griffin for the *Fisherman*, July 25, 1994, shows that Alexandra Morton and other residents living in the Broughton Archipelago noted that wild chum and pink salmon populations had been declining since salmon farms were introduced to their area. (Photo credit: Fisherman Publishing Society)

PANNING DOWN — onto the date: "July 25, 1994."

LETTER — from the Department of Fisheries and Oceans to Alexandra Morton. Date: February 14, 1994.

The subject of the letter pushes forward: Outbreaks of furunculosis in salmon cultured in the Broughton Archipelago.

> MR. MARTLAND (O.S.)
> . . . increased incidence of disease which you saw at the Scott Cove Hatchery, the . . .

ARTICLE — Morton, A.B., and Volpe, J. 2002. A description of escaped farmed Atlantic salmon *Salmo salar* captures and their characteristics in one Pacific salmon fishery area in British Columbia, Canada, in 2000.

> MR. MARTLAND (O.S.)
> . . . capture of escaped Atlantic salmon in commercial fisheries,

HEADLINE — "Farms threaten key gillnet fishery," the *Fisherman*.

> MR. MARTLAND (O.S.)
> . . . and the request by a local fishing lodge to . . .

ARTICLE — Krkosek, M., Ford, J.S., Morton, A.B., Lele, S., Myers, R.A., and Lewis, M.A. 2007. Declining wild salmon populations in relation to parasites from farm salmon.

> MR. MARTLAND (O.S.)
> . . . examine juvenile pink and chum salmon found to be infested with sea lice.

PHOTO — of pink salmon covered in sea lice.

> MR. MARTLAND (O.S.)
> I understand that those,

ON MS. MORTON

> MR. MARTLAND (O.S.)
> . . . and perhaps other factors, led you ultimately to shift your research focus to sea lice in fish farms. Is that a fair capsule description of —

> MS. MORTON
> — Yes. I'd say the original concern had to do with the siting of the fish farms.

> MR. MARTLAND (O.S.)
> Thank you. Since the early 2000s, you've published several academic articles that relate to escaped farmed salmon, sea
> (MORE)

          MR. MARTLAND (O.S.) (CONT'D)
          lice and the effects of fish
          farms on wild pink and chum
          salmon migration routes.

BACK TO MR. MARTLAND

          MR. MARTLAND
          Is that, is that right?

PAN TO MS. MORTON

          MS. MORTON
          I would characterize it as over
          20 scientific papers.

          MR. MARTLAND (O.S.)
          Thank you.

PHOTO — Morton being interviewed at a
Vancouver beach.

          MR. MARTLAND (O.S.)
          You're a popular speaker and
          advocate on issues related
          to protecting wild salmon
          from . . .

PHOTO — Morton is surrounded by supporters at
the Vancouver Art gallery.

          MR. MARTLAND (O.S.)
          . . . potential effects of fish
          farms.

Fisheries
and Oceans

Pêches
et Océans

Pacific Region
Suite 400 - 555 West Hastings St.
Vancouver, B.C.
V6B 5G3

Région du Pacifique
Pièce 400 - 555 rue Hastings ouest
Vancouver (C.-B.)
V6B 5G3

Your file   Votre référence

Our file   Notre référence

FEB 14 1994

Ms. Alexandra Morton
Raincoast Research
Simoom Sound, B.C.
V0P 1S0

Dear Ms. Morton:

OUTBREAKS OF FURUNCULOSIS IN SALMON CULTURED IN THE BROUGHTON ARCHIPELAGO

Thank you for your letter of December 16, 1993 reiterating your concerns regarding outbreaks of furunculosis in salmon cultured in the Broughton Archipelago.

To prevent misunderstandings I will begin with some general comments on furunculosis. This disease is by no means solely a freshwater problem but has been recognized long ago in some strictly marine species, such as sablefish (Anoplopoma fimbria). Independent of any farming or enhancement activity in the area, the causative agent is frequently isolated from wild salmon returning to their rivers of origin. For instance during a survey of wild pinks returning in 1980 to Mathers Creek, Queen Charlotte Islands, our Fish Pathology group isolated Aeromonas salmonicida (As) from 38% of the 50 fish examined. But the disease is also well recognized in the freshwater especially in juveniles.

Despite the name of the disease, furuncles (i.e., boils or ulcerated lesions) are rarely seen. In many outbreaks of furunculosis in juvenile salmon, there are no obvious disease signs. The only telltale sign of the disease occurring can be an increase in losses. Given this information, it is likely that fish in the Scott Cove system have always included some carriers of furunculosis, but it was not recognized as a problem until awareness of the disease increased due to recent outbreaks on the farms.

Drug resistance is a complex subject. One of the important concepts to keep in mind is that drug-resistant mutants arise spontaneously in a population. A drug treatment does not "create" a resistant strain, but simply selects for antibiotic resistant mutants which occur normally as a minor fraction of the bacterial population. Without the selective pressure exerted by the presence of an antimicrobial drug, the resistant strains tend not to persist.

…/2

Canada

Alexandra Morton wrote many letters to government officials at both the provincial and federal levels. This is one of the many responses by DFO trying to downplay her concerns. (Credit: Pat Chamut, Department of Fisheries and Oceans, Canada)

-2-

As you know, five different lysotypes of As are known to occur in B.C. Within each of these lysotype groups there are groups manifesting different antibiotic profiles. The lysotype groups of As found on the farms fall into the five types found in B.C. Therefore there is nothing to indicate that the As found on the farms did not originate in B.C. The antibiotic resistance is likely linked to the antimicrobial treatment regime which the fish had experienced.

While work on establishing the lysotype of the As isolated from adult coho at the Scott Cove Hatchery this fall is not yet complete, it is clear that the strain had a different antibiotic profile from that of the farm fish and is therefore not the same as the strains found on the farms. One half (4) of the As isolates collected from the Scott Cove Hatchery fish by Ms. Dorothee Kieser proved to be oxytetracyline (OTC) resistant, while the other four were sensitive to the drug. All fish had been injected with OTC when entering the hatchery, it is therefore possible that by using OTC we selected for an OTC resistant mutant that these fish were already carrying. The drug resistant strain could also have arisen following their return to the hatchery. Unless uninjected fish from the river are found to be carriers of OTC resistant As, the question of whether the fish carried the resistant strain into the hatchery cannot be answered. Unfortunately, only hatchery fish were available for sampling at the time of the visit.

The impact of As by fish held in net-cages on fish in the surrounding waters is unknown. Of course the transmission of disease agents through the water is a well-recognized route. However, while fish may come in contact with a pathogen, and may, as a result become infected, an actual outbreak of the disease will only occur if the fish are in a susceptible state. In addition, it is difficult to establish where strains isolated from a particular population originated, unless specific marker characteristics are present. In testing wild fish returning to the Kakweikan River we attempted to monitor the possible transfer of As to fish swimming in the vicinity of the net-cages, given that an easily recognized As strain (through its unusual antibiotic profile) had been detected on the farms. As you know the results of this testing were negative.

Our fish health section is planning to undertake further work on the disease interactions between wild and farmed finfish. If strains of these agents recognizable on the basis of their antibiotic resistance profiles show up in the Broughton Archipelago, some of this research would be carried out there.

With respect to your concern that the chinook stock abundance indices are apparently at low levels in the Kingcome and Wakeman River systems, this has been substantiated by the salmon escapement counts for these systems. However, the reason for these low abundances can not be attributed to fish farming activity in the area since similar low escapement numbers have been experienced in the 50's, 60's, 70's and 80's. This is also the case for the chum stock that returns to Viner Sound.

Viner low both in 63 due to slide   ...13
escapement records
Stream reverts Area 12   chums, sockeye
winter springs all time low from   pinks, kings unknown

-3-

Thank you for your interest in these important issues and trust my comments have been helpful.

Yours truly,

P.S. Chamut
Director General
Pacific Region

cc: J. Davis
    R. Ginetz

Farms threaten key gillnet fishery

CONTINUED FROM PAGE 1

B.C. Lands, urging that the fish farm application be turned down.

"The potential lease sites are important gillnet fishing areas and will interfere with the already overcrowded but very important commercial gillnet fishery that occurs in the inlet in July and August of each year," Arkko stated in a June 7 letter to Pentti Leppanen, lands officer for the Cariboo Region where the application was filed.

"Even more important to me is the fear that the migrating salmon will be adversely affected by contact with the farm fish population and effluent," he wrote.

The Fishing Vessel Owners Association had earlier taken the unusual step of writing to B.C. Lands to urge rejection of the application for the five sites, even though its own members would not be directly affected.

"The area is an important migratory route for all salmon species...which gives rise to concerns over disease transmission and possible interactions between native salmon and Atlantic salmon escapees," FVOA executive director Phil Eby stated in a letter to Leppanen.

"The occupancy applications conflict both with salmon migration and the commercial gillnet fishery in Smith Sound, despite the applicant's assertion that the sites applied for are not used for commercial fishing, and we recommend that all of the applications be rejected on that basis," he said.

Leppanen said in an interview that the application had been sent out for comment May 25 to a number of government agencies and interested groups. The company has also been instructed to advertise the application in B.C. Gazette, the Vancouver Sun and the Coast Mountain

News in Bella Coola.

Leppanen noted that the widespread opposition to the fish farm sites "could have a bearing" on what B.C. Lands would do with the application although he added that it would "depend on the nature of that opposition."

Asked about the company's erroneous claim that the area was not used for commercial fishing, Leppanen said that it "was not necessarily erroneous — it could be that he wasn't aware of any commercial fishing in the area."

Leppanen said that 444498 B.C. Ltd. "has a good track record in the Sechelt area" where it is registered. Curiously, although the form filed with B.C. Lands names Kevin Onclin as the principal, the registrar of companies in Victoria lists only one principal and director for 444498 B.C. Ltd. — Roger Engeset.

Applications for land occupancy and use permits are normally processed in about four months, although some

can take as long as a year. The company must also apply to the B.C. Ministry of Agriculture Food and Fisheries for a fish farm permit.

But to grant this application "would effectively displace a lot of people from a spot they've been fishing for many, many years," Arkko said.

The site off Ripon Point as well as the site at the south

end of Dennison Island cut across the entrance to Margaret Bay, preventing fishermen from setting their nets along the shore.

"If we have to keep our nets off the shoreline, it will just ruin the fishery," he said.

Arkko estimated that at least one-third of his annual income comes from the Smith Sound fishery — earnings which would be directly

threatened by the presence of the farms. The area is traditionally opened to gillnetters for two days a week over a three or four-week period, although for the last two years there has been continuous 30-day fishing.

Arkko added that the sites would pose a navigational threat to fishers in the area.

Commercial fishermen complained to the BC Ministry of Agriculture and Fisheries that fish farms were displacing fishermen from traditional fishing locations. The Province still controls the licensing of the fish farm tenures, but a court challenge led by Alexandra Morton found the Province was acting illegally in regulating the farm operations such as stocking densities, etc. (Credit: The Fisherman Publishing Society)

BACK TO MS. MORTON

          MR. MARTLAND (O.S.)
     You've received several
     environmental and conservation
     awards for your advocacy work.

AWARD — Roland Michener conservation award,
"Canadian Wildlife Federation."

PHOTO — Morton receiving her honorary
doctorate.

DOCUMENT — Panning down Morton's honorary
degree.

          MR. MARTLAND (O.S.)
     And in 2010, you were awarded
     an honorary doctorate of
     science from Simon Fraser
     University for your sea lice
     research.

BACK TO MS. MORTON

          MS. MORTON
     That's correct.

          MR. MARTLAND (O.S.)
     Ms. Morton, your name is atop
     a case that we all know about.
     It's sometimes referred to
     as . . .

Alexandra Morton speaks at the Simon Fraser University convocation where she received an honorary doctorate for her work protecting wild Pacific salmon, August 2, 2015. (Photo credit: Anissa Reed)

IN THE SUPREME COURT OF BRITISH COLUMBIA

Citation: *British Columbia (Agriculture and*
              Lands),
              2009 BCSC 136

                                        Date: 20090209
                                        Docket: S083198
                                        Registry: Vancouver

Between:

**Alexandra B. Morton, Pacific Coast Wild Salmon Society, Wilderness Tourism
Association, Southern Area (E) Gillnetters Association, and Fishing Vessel
Owners' Association Of British Columbia**

                                                        Petitioners

And

**Minister of Agriculture and Lands, The Attorney General of British Columbia
on Behalf of The Province Of British Columbia, and
Marine Harvest Canada Inc.**

                                                        Respondents

Before: The Honourable Mr. Justice Hinkson

Reasons for Judgment

Counsel for the Petitioners   Gregory J. McDade, Q.C.
                                                    Lisa C. Glowacki

Counsel for the Respondent Minister of   Nancy E. Brown
Agriculture and Lands and the Attorney   Veronica L. Jackson
General of British Columbia   Cory Bargen a/s

Counsel for the Respondent Marine Harvest   Christopher Harvey, Q.C.
Canada Inc.   Andrew Scarth a/s

Date and Place of Hearing: September 29 and 30, and
                                                October 1 and 2, 2008
                                                     Vancouver, B.C.

Front page of the judgment known as the Hinkson or Morton decision, which overturned a memorandum of understanding, signed September 6, 1988, between the Province of British Columbia and the Federal Government of Canada. This MOU was intended to clarify the role of the two senior levels of government when it came to the new aquaculture industry. It essentially passed any federal role to the Province of BC, which took the lead in licensing and regulation. (See the full MOU in Appendix A.) (Credit: Government of Canada)

DOCUMENT — Front page of Supreme Court of British Columbia ruling.

                    MR. MARTLAND (O.S.)
          . . . the Morton decision
          or the Hinkson decision, but
          it's your name on the style of
          cause.

BACK TO MS. MORTON

                    MR. MARTLAND (O.S.)
          The case states I believe it
          was released in February 2009
          and really changed an important
          part of the regulatory regime,

ON COMMISSIONER COHEN

                    MR. MARTLAND (O.S.)
          . . . putting the DFO largely
          in the driver's seat vis-à-
          vis the regulation of finfish
          aquaculture.

BACK TO MS. MORTON

                    MS. MORTON
          What happened was nobody was
          responsible for the impact of
          the farms to the wild salmon.
          So I didn't make that move
          because I thought the federal
          government would be better. I
          had a long history of problems
                      (MORE)

                    MS. MORTON (CONT'D)
with the federal government
accepting that there were
problems with the industry.
I did it because what was
happening was so wrong, and I
was hoping that in the shake-
up that occurred after, that
it would resettle in a more
logical and beneficial manner
to, to Canada.

**SUPERSCRIPT: Typing. "Cross Examination
Begins . . ."**

                    MR. MARTLAND (O.S.)
Panel members,

ON MR. MARTLAND

                    MR. MARTLAND
. . . thank you very much for
your time with my questions.
I'm only the first of a series
of lawyers.

WIDER

                    MR. MARTLAND
I'll ask next counsel for the
BC Salmon Farmers Association,
Mr. Commissioner, with a
90-minute allocation.

                                    FADE TO:

Juvenile pink salmon covered in sea lice with the scientific name *Caligus clemensi*, a species of lice government and industry refuse to address in salmon farms. Open net pen fish farms are known to cause dramatic increases in sea lice, generally the larger *Lepeophtheirus salmonis*, because they keep thousands of fish, which are hosts for the parasite, in pens together. The dramatic increase in sea lice was described in Scotland as the "sea lice plague." (Photo credit: Alexandra Morton)

Alexandra Morton is interviewed by a local news channel on
Kitsilano Beach, July 16, 2012. (Photo credit: Anissa Reed)

Alexandra Morton surrounded by First Nations women allies during the Cohen Commission
hearings on the Vancouver Art Gallery grounds. The women held a ceremony and sang
the "Women's Song" to help bolster Morton, August 31, 2011. (Photo credit: Anissa Reed)

                    MR. BLAIR
          Good morning, Mr. Commissioner.

**LOWER THIRD: Alan Blair, Counsel, BC Salmon
Farmers Association.**

                    MR. BLAIR
          Members of the panel, for the
          record, my name is Alan Blair.
          I appear as counsel for the BC
          Salmon Farmers Association.

ON MS. MORTON AND MS. STEWART

                    MR. BLAIR (O.S.)
          Ms. Stewart or Ms. Morton,

BACK TO MR. BLAIR

                    MR. BLAIR
          . . . would you like to get in
          and discuss the abstract nature
          of the lack of consensus in
          science? Either one of you can
          lead.

TIGHT ON MS. MORTON

                    MS. MORTON
          The way I see it, the science
          that promotes salmon farming
          is, is what the salmon farmers
                    (MORE)

MS. MORTON (CONT'D)
use, but science that I have
done has often not been even
cited in their responses to me.
There's been a lot of debate
back and forth about my method,
so that stimulated me to, for
example, when I did a study
on do sea lice kill juvenile
pink salmon, I invited Dr.
Brian Riddell and Dr. Brent
Hargreaves to come and view
the experiment right off the
bat so that I didn't need to
meet their opposition later.
I could adapt the study right
away, which I did. And yet,
when Dr. Jones finds a highly
conflicting result in that pink
salmon are completely resistant
to sea lice after they weigh
0.7 grams, he will not even
address the difference between
our findings. So, for me, I
watched the fish die, I'm an
eyewitness to it. I measured
it and I went the lengths of
putting it into a journal, so
it's, it's very difficult. I
don't work on sea lice any more
because I figured it out. Where
there's fish farms, there's sea
lice.

PANNING ACROSS THE GALLERY

> MR. BLAIR (O.S.)
> Ms. Morton, there's fish
> everywhere in the ocean and
> there's sea lice, not just
> with fish farms.

ON MR. BLAIR

> MR. BLAIR
> You agree that sea lice are a
> naturally occurring phenomenon
> and there's lots of sea lice in
> areas where there are no fish
> farms, surely?

BACK TO MS. MORTON

> MS. MORTON
> Yes, the sea louse is a benign
> crustacean parasite and —

> MR. BLAIR (O.S.)
> — And you agree that salmon
> farms come . . .

BACK TO MR. BLAIR

> MR. BLAIR
> . . . into the ocean, to the
> marine environment devoid of
> sea lice because they come from
> freshwater hatcheries and they
> come in without sea lice?

Another example of juvenile salmon with sea lice on them. The dark patches on the fish are where the sea lice have chewed through the skin, causing scars. The black dots are holes in the skin caused by the large female lice. The two large sea lice are known as the salmon louse (*Lepeophtheirus salmonis*). The dark line down their centre indicates they are feasting on the fish's blood. Normally these lice are found on adult salmon, which are protected by scales. (Photo credit: Alexandra Morton)

An early fish farm in BC. Early farms were relatively small and were usually owned by small BC companies or sole proprietors. (Photo credit: Fisherman Publishing Society)

The aquaculture industry in BC was taken over by large multinationals, and the size of the farms grew significantly. (Photo credit: Alexandra Morton)

                    MS. MORTON (O.S.)
          Absolutely.

                    MR. BLAIR
          So they pick up the sea lice
          from the wild fish?

BACK TO MS. MORTON

                    MS. MORTON
          That's right.

                    MR. BLAIR (O.S.)
          And wild fish would have sea
          lice whether the salmon farms
          were there or not?

BACK TO MS. MORTON

                    MS. MORTON
          There's two, for me, profound
          issues with salmon farms. One
          is amplification of local
          endemic parasites and diseases,
          and the other is exotic. So the
          sea lice fall into a dangerous
          amplification because all the
          Pacific salmon that come into
          the coast in the fall die,
          which is a remarkable natural
          thing. Why does nature do that?
          It's to break the cycle of
          disease, but salmon farms have
          given the sea lice a place to
                    (MORE)

>   MS. MORTON (CONT'D)
> over-winter and, and reproduce
> and so the young fish meet them
> before they're ready.

PAN TO MR. BLAIR

>   MR. BLAIR
> I'm sorry. Mature salmon come
> back into the waters of BC and
> they die to break the cycle of
> disease? You figured that out,
> that's why they die?

BACK TO MS. MORTON

>   MS. MORTON
> I wish I'd figured it out, but
> no, it wasn't me.

BACK TO MR. BLAIR

>   MR. BLAIR
> Okay. So a bit of an
> extrapolation there, yes?

>   MS. MORTON (O.S.)
> No.

>   MR. BLAIR
> I'm going to suggest to you
> that your experience with sea
> lice gained some notoriety as a
> result of some studies you did
>   (MORE)

                    MR. BLAIR (CONT'D)
          in the Broughton Archipelago in
          the early 2000s. Do you agree
          with that?

PAN TO MS. MORTON

                    MS. MORTON
          It gave me a doctor of science
          at the Simon Fraser University.

PAN BACK TO MR. BLAIR

                    MR. BLAIR
          Honorary doctor?

                    MS. MORTON (O.S.)
          Correct.

                    MR. BLAIR
          Yes. And you'll agree that from
          those first reports, there have
          been, it's like a Ping-Pong
          match, or a tennis match, there
          have been reports in support
          of those earlier reports, and
          reports that are quite, quite,
          ah, scathing in terms of the
          methodologies used? You'll
          agree that it's a Ping-Pong
          match, or a tennis match back
          and . . .

PANNING BACK TO MS. MORTON

                    MR.BLAIR (O.S.)
. . . forth on some of that
science? You'll agree you've
certainly read it, correct?

ON MS. MORTON

                    MS. MORTON
I would characterize it more as
mud-slinging. I don't see the
Ping-Pong. We all agree, those
of us that are out there, that
the sea lice are coming from
the farms. Even Dr. Marty's
study said as the number of
sea lice increase on the farms,
they increase on the adjacent
pink and chum salmon. That
should have put the whole
argument to rest right there.

                    MR. BLAIR (O.S.)
Is the mud-slinging, from your
opinion, only coming from one
direction?

ON MR. BLAIR

                    MR. BLAIR
Is it only the salmon farmers
who are flinging mud, or a
little bit of mud going both
ways?

BACK TO MS. MORTON

                    MS. MORTON
          I'm defending myself at this
          point.

                    MR. BLAIR (O.S.)
          I'm sorry?

                    MS. MORTON
          I'm defending myself at this
          point. I'm not, ah, I'm not
          going to just quietly take it,
          because it needs to be argued
          back.

BACK TO MR. BLAIR

                    MR. BLAIR
          But my question, the question
          started around diverse science
          and whether or not there's a
          conflict in science or whether
          there's a consensus. It's
          clear there's no consensus.
          My question of you is, do
          you agree that on both sides
          of that equation, that the
          scientific debate, parties are
          coming to different conclusions
          for different reasons? Do you
          agree with that?

WHIP BACK TO MS. MORTON

                    MS. MORTON
          Definitely for different
          reasons, but the biology of it,
          Mr. Blair, is extremely easy,
          and whether you're talking to a
          scientist in Norway, Scotland,
          Ireland, Chile, Eastern Canada
          or British Columbia, because
          I talk to them all, fish farms
          definitely amplify sea lice,
          and we have got to move past
          that.

PAN BACK TO MR. BLAIR

                    MR. BLAIR
          And so, all of the reports
          that would disagree with that
          position of yours you say are
          categorically wrong?

PAN TO MS. MORTON

                    MS. MORTON
          They do not disagree with
          that position. It's their
          interpretation.

                    MR. BLAIR (O.S.)
          I see. Ms. Stewart, your turn.

ON MR. BLAIR

                    MR. BLAIR
          Let's talk about good faith.

POP-UP — CAAR: Coastal Alliance for Aquaculture Reform

The people in the gallery groan.

> MR. BLAIR
> Certainly, CAAR enters the
> dialogue in good faith. CAAR
> intends to be productive and
> constructive in dialogue with
> multi stakeholders, yes?

> MS. STEWART
> That was our intent, yes.

> MR. BLAIR (O.S.)
> And I'm assuming you haven't
> abandoned that attempt?
> You're still trying to be
> there in good faith and have
> collaborative dialogue?

BACK TO MS. STEWART

She tries to speak, but is interrupted by Mr.
Blair.

> MR. BLAIR (O.S.)
> I'm not suggesting otherwise,
> by the way.

**LOWER THIRD: Catherine Stewart, Former
Campaign Director, Living Oceans Society.**

                    MS. STEWART
          No, we haven't abandoned
          the dialogue, but I would
          say that there is a pall of
          discouragement over many of the
          member groups of CAAR at the
          lack of progress and the, the
          glacial pace of implementation.

                    MR. BLAIR (O.S.)
          You're not alleging bad faith
          of any of the other parties,
          you're just suggesting that
          it's a complex area to reach
          consensus . . .

Ms. Stewart expresses her disagreement as Mr.
Blair continues to speak.

ON MR. BLAIR

                    MR. BLAIR
          . . . or are you alleging bad
          faith?

PANNING TO MS. STEWART

                    MS. STEWART (O.S.)
          I'm not alleging bad faith.

ON MS. STEWART

                    MS. STEWART
But I would say that, that to
a degree, it's my belief, and I
certainly can't speak for the
other groups of CAAR, or even
for Living Oceans on this, this
is my personal perspective,
that I think that there has
been some mastery of the art of
foot-dragging.

                                    DISSOLVE TO:

                    MR. MARTLAND
Mr. Commissioner, the counsel
for the Province with 30
minutes.

Mr. Prowse stands up and approaches the
microphone.

                    MR. PROWSE
    Yes.

**LOWER THIRD: Clifton Prowse, QC, Senior
Counsel, Province of British Columbia.**

                    MR. PROWSE
Mr. Commissioner, Cliff Prowse
for the Province of British
Columbia.

ON MS. MORTON

>MR. PROWSE (O.S.)
>For you, Ms. Morton, and . . .

ON MR. PROWSE

>MR. PROWSE
>. . . it has to do with your
>undergraduate education. So
>I, I concede that it's not
>an important topic, perhaps,
>but it's one that's piqued
>curiosity. You have a
>bachelor's degree in science
>from 1977?

PAN TO MS. MORTON

>MS. MORTON
>Yes, I do.

>MR. PROWSE (O.S.)
>And what was the discipline?

>MS. MORTON
>Interdisciplinary.

BACK TO MR. PROWSE

>MR. PROWSE
>And did you do a thesis?

>MS. MORTON (O.S.)
>No, I did not.

                    MR. PROWSE
You don't have a master's
degree or a PhD or veterinary
degree or pathology specialty.
Is that correct?

PAN TO MS. MORTON

                    MS. MORTON
That's correct. I simply have
an honorary doctorate.

BACK TO MR. PROWSE

                    MR. PROWSE
And you don't have an advanced
degree in mathematics or
epidemiology?

BACK TO MS. MORTON

                    MS. MORTON
     No.

ON MR. PROWSE

                    MR. PROWSE
As a registered professional
biologist, you're under an
ethical obligation to undertake
only those assignments for
which you are qualified?

PAN TO MS. MORTON

MS. MORTON
I don't have assignments. I
just have a personal interest.

BACK TO MR. PROWSE

MR. PROWSE
So, as a registered
professional biologist,
does that mean that you can
undertake any assignments,
whether they're qualified or
not?

OVER TO MS. MORTON

MS. MORTON
When, when you do a scientific
study, the bar that you pass is
whether it's accepted by the
journal. And what the journal
does when you do a scientific
article is they send it out
to the people they think are
going to oppose you. And if it
passes review with the journal,
you really need to take this
up with the journals who have
published me, the *ICES Journal
of Marine Science*, the *Journal
of Science*, the *American
Fisheries Journal Transactions*.
You could go on on this point
forever, but that's the bar,
just so as you know.

BACK TO MR. PROWSE

> MR. PROWSE
> All right. You have published
> peer-reviewed research, as
> you've just told us?

> MS. MORTON (O.S.)
> Yes, that's correct.

> MR. PROWSE
> With respect to the peer-review
> process, isn't the purpose
> of the journal to review it
> generally with your, with
> peers, rather than finding
> people that are going to
> oppose?

PAN BACK TO MS. MORTON

> MS. MORTON
> Oh no, quite to the contrary.
> They want, they don't want to
> make the error of publishing
> something that's wrong or
> political or for purposes other
> than the science itself, so for
> example, um, Dick Beamish's
> paper was sent to me. So we,
> we, we publicly held opposing
> opinions and they sent it to
> me. So, no, they're looking for
> a broad opinion.

BACK TO MR. PROWSE

           MR. PROWSE
All right. Ms. Parker, did you
have a comment?

PAN TO MS. PARKER

           MS. PARKER
I just wanted to, I will
agree with Ms. Morton's last
statement, that they're
looking, that during peer
review, it's about having a
broad, a broad perspective, but
I would say that looking for
opponents is inconsistent with
the academic integrity of peer
review.

BACK TO MR. PROWSE

           MR. PROWSE
Did you want to respond to
that, Ms. Morton?

TO MS. MORTON

           MS. MORTON
No, not really.

BACK TO MR. PROWSE

                    MR. PROWSE
     You agree that research
     involves the generation of
     hypotheses?

OVER TO MS. MORTON

                    MS. MORTON
     It's the testing of hypotheses.
     So you start with a hypothesis
     and then you go out and you try
     to understand the validity of
     it as best you can.

PAN TO MR. PROWSE

                    MR. PROWSE
     And —

                MS. PARKER (O.S.)
     — Excuse me.

PAN TO MS. PARKER

                    MS. PARKER
     Research begins with a null
     hypothesis and then you go out
     and try to disprove it.

BACK TO MR. PROWSE

                    MR. PROWSE
     Do you want to respond to that,
     Ms. Morton?

OVER TO MS. MORTON

                    MS. MORTON
          No.

OVER TO MR. PROWSE

                    MR. PROWSE
          Now, as we've seen in this
          Commission, there may be peer-
          reviewed responses and peer-
          reviewed counter-responses?

BACK TO MS. MORTON

                    MS. MORTON
          Yes, that's correct.

                    MR. PROWSE (O.S.)
          And as a scientist, you are
          aware there's a significant
          difference between hypothesis
          and proof?

                    MS. MORTON
          Yes.

                    MR. PROWSE (O.S.)
          And you'd agree that peer-
          reviewed research is generally
          entitled to more weight than
          other scientific articles which
          have not been peer-reviewed?

                    MS. MORTON
          Yes, that's correct.

PAN TO MR. PROWSE

                    MR. PROWSE
          And you published with Dr.
          Larry Dill?

                    MS. MORTON (O.S.)
          Yes, I've co-published with
          him.

DOCUMENT — Exhibit 1540, Impacts of salmon
farms on Fraser River sockeye salmon: Results
of the Dill Investigation.

                    MR. PROWSE (O.S.)
          Now, Mr. Lunn, could we have
          Exhibit 1540, summary page 34
          in the ordinary numbering?

ON PAGE 34 — We see the summary.

                    MR. PROWSE (O.S.)
          So in the end of the first
          paragraph at Exhibit 1540,

BACK TO MR. PROWSE

                    MR. PROWSE
          . . . Dr. Dill attempts to
          narrow the issue for the
          Commission with respect to the
                    (MORE)

                    MR. PROWSE (CONT'D)
          cause of the long-term decline
          in the, especially returns in
          2009. Um, and he specifically
          says that there's no evidence
          to support the following items
          with respect to those declines,
          namely lice, benthic and
          pelagic impacts and escapes,
          and he says that they're
          none of them "are likely to
          be sufficient, alone or in
          concert, to cause either the
          long-term population declines
          or the especially low returns
          in 2009." Do you agree with
          that statement?

PAN TO MS. MORTON

                    MS. MORTON
          I would add to that list
          pathogens and then lice and
          pathogens in concert could be a
          large factor in the declines.

BACK TO MR. PROWSE

                    MR. PROWSE
          All right. So you agree with
          the statement as written but
          you would add the factor of
          pathogens?

BACK TO MS. MORTON

                    MS. MORTON
        Yes. Because lice are such, a,
        a, an effective vector and so,
        yeah, they definitely, in my
        mind, play a role because they
        move between the farmed fish
        and the wild fish.

BACK TO MR. PROWSE

                    MR. PROWSE
        All right. Now, until the year
        2000 your publications were
        largely on killer whales?

BACK TO MS. MORTON

                    MS. MORTON
        Yes, they were. And dolphins.

BACK TO MR. PROWSE

                    MR. PROWSE
        Mr. Lunn, could we have Exhibit
        1557, please?

DOCUMENT — Exhibit 1557, Sea lice dispersion
and salmon survival in relation to salmon
farm activity in the Broughton Archipelago.

                    MR. PROWSE (O.S.)
        I want to turn to the question
        of fallowing.

PUSH IN — on the authors listed, with
emphasis on Morton's name.

> MR. PROWSE (O.S.)
> This is a paper that you were
> a co-author of with respect to
> sea lice dispersion and salmon
> survival in relation to salmon
> farm activity in the Broughton
> Archipelago?

> MS. MORTON (O.S.)
> Yes, that's correct.

PAGE TURN — to page 155.

> MR. PROWSE (O.S.)
> This is on the left-hand side
> of the page. The statement
> says, "Based on escapement
> data" there were . . .

THE WORDS — "there were no significant
differences in survival" are highlighted.

> MR. PROWSE (O.S.)
> . . . "no significant
> differences in survival that
> corresponded to sea louse
> abundance . . .

BACK TO MR. PROWSE

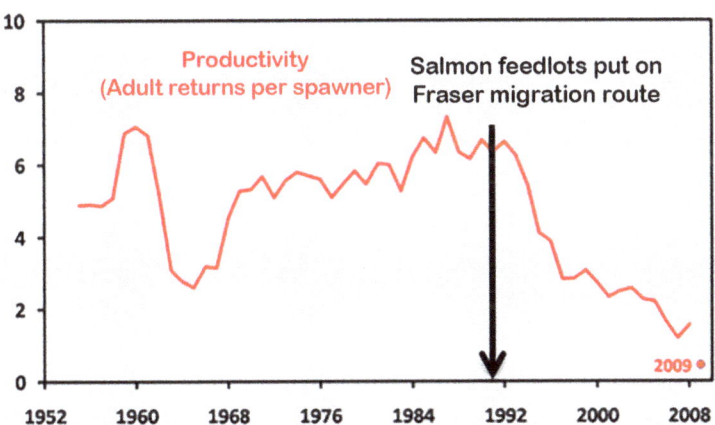

This graph was an exhibit at the Cohen Inquiry. It tracks the number of adult sockeye salmon that return per pair of spawners. It reveals that historically, there were six adult fish returns per spawner, but after the introduction of open net pen fish farms, the recruitment number began to drop. By 2009, the number was below the critical replacement number of two. (Black text on graph added by Alexandra Morton.)
(Credit: Cohen, B. October 2012. *The Uncertain Future of Fraser River Sockeye*, Volume 2.)

                    MR. PROWSE
. . . and juvenile salmon
mortality on the migration
route containing active
farms relative to unexposed
populations north of the
Broughton Archipelago."

PAN TO MS. MORTON

                    MS. MORTON
Yes, that's correct.

                    MR. PROWSE (O.S.)
And then on page 149 under
the heading "Escapement and
Survival Analysis."

ON MR. PROWSE

                    MR. PROWSE
So the first sentence there:
"Survival among rivers, based
on escapement data, was highly
variable, and there was no
detectable difference in mean
survival for the Broughton
Archipelago relative to the
central coast. . . . only the
Embly River clearly corresponds
to the fallow migration route.
That population experienced
very poor survival, with a 90%
decline, although it was
                    (MORE)

>   MR. PROWSE (CONT'D)
>   subject to fallow
>   intervention." So that, is it
>   correct?

PAN TO MS. MORTON

>   MS. MORTON
>   Yes. And I, I really appreciate
>   you bringing up this paper,
>   because this speaks to the
>   integrity of my work. I found
>   a finding here that runs
>   contrary to what I generally
>   have found and put out, but you
>   need to understand that when I
>   began studying sea lice, the
>   salmon farms were not treating
>   prophylactically. They were not
>   treating to protect the pink
>   salmon and the chum salmon of
>   the Broughton and the average
>   number of lice was 11. And in
>   the years after that, it was
>   still extremely high. By the
>   time I did this work, which
>   included 87 plankton tows in
>   the dead of winter, 20 minutes
>   for each, I looked at 9,000
>   fish, live, in the months
>   between March and um, and um,
>   May, and the average number of
>   lice was 0.3. And so what the
>   farms had done is they had used
>   (MORE)

                    MS. MORTON (CONT'D)
a chemical to drive the lice
numbers down. If I felt that
that chemical was going to work
forever on lice and if I felt
that was the only problem with
salmon farms, then I would
be relieved and be able to go
back to studying whales. But
this paper should bring to this
court the fact that when I find
something that does not support
my basic belief about this
industry, I will publish it as
well.

PAN TO MR. PROWSE

                    MR. PROWSE
So in your peer-reviewed
publications you're making
a, research indicating that
fallowing did not have any
effect on wild salmon survival
under that —

BACK TO MS. MORTON

                    MS. MORTON
— Now you are cherry-picking,
because the previous paper I
did on fallowing, which looked
at the years 2003, '02 and '04,
I found a different result
                    (MORE)

                    MS. MORTON (CONT'D)
          because at that time the lice
          were not being controlled by
          the salmon farming industry and
          the fallowing, the removal of
          the fish, not only dropped the
          number of lice enormously, but
          Dr. Beamish published on the
          year class that was treated
          to the fallow and those pink
          salmon survived better than
          in the history ever of pink
          salmon. So you really just, you
          can't latch onto one detail.
          There was an enormous amount of
          drugs used to accomplish this
          result.

ON COMMISSIONER COHEN — He is listening
intently.

OVER TO MR. BACKMAN

                    MR. BACKMAN
          I think it's important to bring
          up a point, if I may, that,
          you know, Ms. Morton has made
          several recommendations, several
          references to the paper with
          Dr. Beamish here that was done
          looking at the 2004 . . .

**LOWER THIRD: Clare Backman, Director of
Environmental Compliance and Community
Relations, Marine Harvest.**

                    MR. BACKMAN
. . . return of salmon to
the archipelago, and from
perspectives, my perspective on
this in my speaking with, uh,
with Dr. Beamish was that yes,
some wording was added to that
document but the thrust of that
document is about changes in
the regime of the ecosystem of,
of the Broughton Archipelago.
And, uh, it's consistently
misrepresented here that it's
all about whether some farms
were operating or not. Um, and
it's true that Ms. Morton was
a peer reviewer. I think it's
also true that, ah, the peer-
review process went on a very,
very long time and was finally,
at very great length between
the two of them before it could
be worked out that he would add
a few more words to the credit,
that there were some farms
that were operating, some that
weren't. But the focus of his
work was on the regime change
and the ecosystem change in the
Broughton Archipelago.

BACK TO MS. MORTON

MS. MORTON
Okay. I was not going to
discuss this, but this has to
be brought up now. Dr. Beamish,
for a period of months —

MR. PROWSE (O.S.)
— Well, I'm sorry, Dr. Morton —

MS. MORTON
— refused to acknowledge that
the farms —

MR. PROWSE (O.S.)
— Sorry, Mr. Commissioner, I
would like to —

MS. MORTON
— on the fallow route were
empty.

MR. PROWSE (O.S.)
I would like to move on. I
obtained the —

MS. MORTON
— And the words that were added
were to say —

MR. PROWSE (O.S.)
— I didn't —

MS. MORTON
— that those farms were empty.

                    MR. PROWSE (O.S.)
          I didn't ask this question.
          I didn't ask for the
          intervention. I'm not asking
          for a response. I'd like to
          move on, Mr. Commissioner.

                    MS. MORTON
          I'm sure.

ON MR. PROWSE

                    MR. PROWSE
          Dr. Morton, um, in addition to
          doing your scientific research,
          you also campaign publicly?

PAN TO MS. MORTON

                    MS. MORTON
          After doing 10 years of
          research I began to campaign
          publicly.

PAN TO MR. PROWSE

                    MR. PROWSE
          Starting when are you
          identifying your campaigning?

PAN BACK TO MS. MORTON

When the government ignored first her letters and then her research on the impacts of open net pen fish farms, Alexandra Morton decided to take her concerns directly to the public. She went from concerned citizen to scientist to activist. This photo was taken at the Vancouver Art Gallery during one of several rallies held in support of wild salmon. (Photo credit: Anissa Reed)

                    MS. MORTON
          Well, it depends how you define
          public campaigning, 'cause it
          started with 10,000 letters to
          DFO and then it went into doing
          10 years of research on sea
          lice. There was engaging in the
          salmon aquaculture dialogue and
          review and the CRIS study . . .

**POP-UP — CRIS Study: Coastal Resources
Inventory Studies**

                    MS. MORTON
          . . . and the special
          legislative committee, so
          there's been a lot of
          participation in public
          processes, and when my, when
          I began to see that the
          archipelago that I was living
          in was still suffering from
          this industry, I figured that
          the next step was to go to the
          public and so that, you know,
          that really got started about
          two years ago.

PAN TO MR. PROWSE

                    MR. PROWSE
          As a scientist, when you
          speak publicly, do you find it
          necessary to simplify complex
          issues?

63.

BACK TO MS. MORTON

                MS. MORTON
    Yes, I do.

                MR. PROWSE (O.S.)
    And as a campaigner, it's
    important to you to get your
    message out and to communicate
    effectively?

                MS. MORTON
    As someone who's trying to
    protect her home, yes, I
    do find it's helpful to
    communicate clearly.

OVER TO MR. PROWSE

                MR. PROWSE
    So you not only simplify, but
    you present your message in the
    most effective way?

BACK TO MS. MORTON

                MS. MORTON
    I like to communicate things as
    clearly as possible.

PAN TO MR. PROWSE

                    MR. PROWSE
          When you present as a
          campaigner and not as a
          biologist, you do not have
          to confine yourself to your
          expertise?

BACK TO MS. MORTON

                    MS. MORTON
          The biologist is underlying
          everything. If, if the
          government had reacted to my
          concerns, I would never be
          talking publicly.

PAN BACK TO MR. PROWSE

                    MR. PROWSE
          And in fact, campaigners have
          great freedom in what they say
          to media?

BACK TO MS. MORTON

                    MS. MORTON
          There's nobody restraining
          my freedom. I'm not paid
          by anybody, so I try to
          communicate as clearly and as
          fairly as I see possible.

PAN BACK TO MR. PROWSE

                    MR. PROWSE
          And if you have to choose
          between clear and fair, what
          choice do you make?

OVER TO MS. MORTON

The spectators in the gallery mutter their
disapproval at the line of questioning, and
a lone "boo" can be heard from the restless
crowd.

                    MS. MORTON
          I choose fair as often as
          possible, yes.

BACK TO MR. PROWSE

                    MR. PROWSE
          Effective media statements
          encourage simple, startling
          messages?

PAN TO MS. MORTON

                    MS. MORTON
          The issue is startling,
          and clear is required to
          communicate it.

BACK TO MR. PROWSE

                    MR. PROWSE
          So media messages do not
          involve the peer-review
          processes that restrict what
          scientists say in peer-reviewed
          literature?

PAN TO MS. MORTON

                    MS. MORTON
          The media messages that I use
          are based on my experience and
          peer-reviewed science.

ON MR. PROWSE

                    MR. PROWSE
          And you do not have to follow
          governmental restrictions
          preventing you from talking
          to the media about Commission
          matters?

BACK TO MS. MORTON

                    MS. MORTON
          No, I'm, I, I can say whatever
          I feel is right to say.

PAN TO MR. PROWSE

                    MR. PROWSE
     And since you're not a
     veterinarian, you do not have
     to be restricted by obligations
     of veterinarian ethics?

BACK TO MS. MORTON

                    MS. MORTON
     I am not a veterinarian.

PAN TO MR. PROWSE

                    MR. PROWSE
     And you feel free to attack
     those who disagree with you?

                    MS. MORTON (O.S.)
     I feel free to defend the home
     that I love and want to see
     thrive.

BACK TO MR. PROWSE

                    MR. PROWSE
     Would you not agree that,
     at the end of the day, the
     Commission should place more
     reliance on your peer-reviewed
     publications than on your
     statements going beyond your
     field of expertise?

PAN TO MS. MORTON

> MS. MORTON
> No, I disagree. He's faced with
> an enormous task of weighing
> this evidence out and I don't
> envy the job.

FADE TO:

PAN TO MR. TAYLOR

> MR. TAYLOR
> Mitchell Taylor, and with
> me is Jonah Spiegelman, Mr.
> Commissioner.

ON COMMISSIONER COHEN

> MR. TAYLOR (O.S.)
> I'm going to start my questions,
> if I may,

ON MR. TAYLOR

> MR. TAYLOR
> . . . directing questions to Ms.
> Stewart and Ms. Morton, and I'm
> going to ask questions about
> jurisdiction and regulation. I
> understand that both of you are
> unsatisfied or dissatisfied
> with the provincial regulation
> and with the federal regulation
> (MORE)

> MR. TAYLOR (CONT'D)
> of aquaculture. Provincial
> before December of 2010, and
> the federal afterwards, of
> course. And in that, I think we
> all recognize that at all times
> both governments have had a
> role in aquaculture. It's where
> the majority of the regulatory
> power is, or regulatory
> authority. Now, Ms. Stewart,
> will you agree with me that one
> or the other of the provincial
> or federal government is going
> to have to be the regulator?

PAN TO MS. STEWART

> MS. STEWART
> Yes.

OVER TO MS. MORTON

> MR. TAYLOR (O.S.)
> And, Ms. Morton, you agree with
> that?

> MS. MORTON
> Ah, yes.

PAN TO MR. TAYLOR

> MR. TAYLOR
> All right. So far, so good.

                    MS. STEWART (O.S.)
          Maybe we should end there.

                    MR. TAYLOR
          No, we won't. Thank you for
          that offer, though. Now, with
          that, do you accept that open
          net pen salmon aquaculture
          is an activity that can occur
          somewhere on the British
          Columbia coast?

PAN TO MS. STEWART

                    MR. TAYLOR (O.S.)
          Ms. Stewart?

                    MS. STEWART
          I think it would be difficult
          to argue that one open net
          pen in an area the size of the
          Broughton Archipelago would
          be, you know, fundamentally
          destructive to ocean
          ecosystems. The, the question
          has to be one of scale. Um,
          you know, the current numbers
          of open net cages we have, I
          believe, are unsustainable, and
          I don't believe that we can, I
          certainly don't believe that we
          can increase production. The
          burden on the environment is
          already too high.

>   MR. TAYLOR (O.S.)
>   Are there places or locations
>   on the BC coast that, in your
>   view, can have sustainable
>   aquaculture coexisting with
>   wild stocks . . .

PAN TO MR. TAYLOR

>   MR. TAYLOR
>   . . . where you have multiple
>   farms and, whatever that number
>   is, multiple farms that are
>   supporting an industry similar
>   to what we have now, are there
>   locations where that can occur?

BACK TO MS. STEWART

>   MS. STEWART
>   Sure, in closed containment
>   systems.

>   MR. TAYLOR (O.S.)
>   No, no, we're talking about
>   open net.

>   MS. STEWART
>   Again, no, I don't think so.
>   Not multiple farms, no. The
>   impacts and the weight of
>   evidence suggests that the
>   impacts are already too great
>   and the risks are extremely
>   high.

PAN TO MR. TAYLOR

                    MR. TAYLOR
          No location anywhere on the
          coast?

                    MS. STEWART (O.S.)
          Well, there's over 9,000
          individual salmon stocks on the
          coast of BC.

BACK TO MS. STEWART

                    MS. STEWART
          Virtually the entire coast is
          in a migratory route for wild
          salmon, and wild salmon are
          really the foundation of the
          coastal Pacific ecosystem.
          They feed a variety of species
          as we know, multiple species,
          over 300. They even feed the
          forests.

PAN TO MR. TAYLOR

                    MR. TAYLOR
          All right. I have your point
          on that, thank you. Ms. Morton,
          what do you say?

TO MS. MORTON

                    MS. MORTON
No, there's no place that open
net pens can coexist with wild
fish.

                   MR. TAYLOR (O.S.)
Now, your point, Ms. Morton,
has been, as I understand it,
that the problem is the salmon
farms on the migratory routes.
Is that your point?

                    MS. MORTON
The problem is that salmon
farms amplify pathogens, they
break the natural laws, and so
they disrupt the ecosystem that
they're in.

PAN TO MR. TAYLOR

                    MR. TAYLOR
And so you say that even if
they weren't close to where
the salmon are migrating, it's
still a problem?

BACK TO MS. MORTON

                    MS. MORTON
If there was such a place in
British Columbia, it wouldn't
be a problem, but I've spent
                    (MORE)

                    MS. MORTON (CONT'D)
a long time looking for that
place, and it doesn't appear to
exist.

BACK TO MR. TAYLOR

                    MR. TAYLOR
All right. I have your
point, thank you. Now, Ms.
Morton, you took steps to
bring before the courts of
British Columbia the question
of whether the provincial
or federal government have
jurisdiction over aquaculture,
and ultimately you obtained
a judgment that for finfish
aquaculture, it is the federal
government who has regulatory
control. We all know that. Now,
am I correct that you went into
that litigation knowing that
you might win and the court
might say that it's a matter of
federal jurisdiction?

PAN TO MS. MORTON

                    MS. MORTON
Yes, based on my lawyer,
Gregory McDade, we thought we
would win.

                    MR. TAYLOR (O.S.)
          All right. And so here we are,
          and you did win, and there is
          federal jurisdiction.

OVER TO MR. TAYLOR

                    MR. TAYLOR
          So, when you pursued that
          litigation and ultimately were
          successful, did you, as you
          were pursuing the litigation,
          have an understanding that if
          you were successful and there
          was federal jurisdiction, there
          would have to be a fairly quick
          turnover from provincial to
          federal regulatory regime?

PAN TO MS. MORTON

                    MS. MORTON
          Yes, an understanding existed.

                    MR. TAYLOR (O.S.)
          And you full well knew the
          complexity of the subject
          matter. You've spoken a lot
          about that, is that right?

                    MS. MORTON
          For me it's very clear, the
          DFO . . .

POP-UP — DFO: Department of Fisheries and Oceans

> MS. MORTON
> . . . needs to protect the wild
> salmon from whatever it is, so
> the complexities of regulation
> and the bureaucracy behind
> it, I don't fully grasp, but
> I do grasp the biology of the
> situation.

PAN TO MR. TAYLOR

> MR. TAYLOR
> Well, I suppose you might say
> that it's all very simple if
> everyone agreed with you, but
> you know that everyone doesn't
> agree with you, correct?

BACK TO MS. MORTON

> MS. MORTON
> It's really not a matter of
> agreeing with me. It's a matter
> of an honest appraisal of
> the natural world and what's
> happening, the dynamics between
> the two populations.

PAN TO MR. TAYLOR

                    MR. TAYLOR
All right. That's your view
of it, I take it, but you full
well know that there are people
who hold contrary views to what
you do, correct?

BACK TO MS. MORTON

                    MS. MORTON
I full well do, but I believe
they're wrong.

BACK TO MR. TAYLOR

                    MR. TAYLOR
Yes, I know that. And many of
those people that hold contrary
views are very respected
scientists, correct?

PAN TO MS. MORTON

                    MS. MORTON
Um. Are they respected? I have
honestly lost a lot of respect
in this process, I have to
be honest with you. I don't
mean to be harrasive with that
statement, but when you are
looking at the fish and you
have put enormous effort into
it, there is, it's inescapable,
the effect of this industry,
                    (MORE)

                    MS. MORTON (CONT'D)
          whether it's toxic algae
          blooms, displacement of the
          whales —

                    MR. TAYLOR (O.S.)
          — Yes, we're talking about —

                    MS. MORTON
          — lice, bulging eyeballs —

                    MR. TAYLOR (O.S.)
          — respected scientists —

                    MS. MORTON
          — blackening skin. It's just
          really evident —

                    MR. TAYLOR (O.S.)
          — Ms. Morton, we're talking
          about respected scientists
          at the moment, if we could
          stick to the question, please.
          You know that people like Dr.
          Noakes, Dr. Beamish, Dr. Jones,

PAN TO MR. TAYLOR

                    MR. TAYLOR
          . . . Dr. Johnson, Dr. Dill,
          with certain caveats, and
          others, all have a different
          view than you, don't you?

PAN TO MS. MORTON

                    MS. MORTON
I don't believe Dr. Dill does
have a different view from me,
but the rest —

                    MR. TAYLOR (O.S.)
— Well —

                    MS. MORTON
— of them I know have a very
different view.

BACK TO MR. TAYLOR

                    MR. TAYLOR
Now, with all that, do you
accept that there is complexity
— I think you do — to creating
a new regulatory regime?

BACK TO MS. MORTON

                    MS. MORTON
I accept there are complexities.

PAN TO MR. TAYLOR

                    MR. TAYLOR
And there also has to be
consultation with quite a
number of interested parties
and stakeholders, including
First Nations, doesn't there?

PAN TO MS. MORTON

                    MS. MORTON
Yes.

                    MR. TAYLOR (O.S.)
And will you agree with me that
in the time available, which is
approximately 12 to 16 months
from the decision until the
regulatory regime came into
play, that a lot of good work
was done in order to set up a
new regulatory regime?

                    MS. MORTON
No, I don't think it was good
work.

BACK TO MR. TAYLOR

                    MR. TAYLOR
All right.

                    MS. STEWART (O.S.)
I would like to make a
comment on that, actually,
because I was engaged in
discussions . . .

PAN TO MS. STEWART

                    MS. STEWART
. . . with both the provincial
and federal governments after
the Morton decision came down,
                (MORE)

                    MS. STEWART (CONT'D)
and for the first six months
there was an awful lot of
debate around what they were
going to do. Um, the, the
province wasn't sure they
wanted to totally relinquish
control, the federal government
wasn't sure they wanted to
completely accept control. I
sat in meetings with Trevor
Swerdfager and Harvey Sasaki
from the provincial government,
where they were debating
whether or not they would
negotiate a constitutionally
acceptable memorandum of
understanding around sharing
jurisdiction. There was a long
debate, and it wasn't until
around September or October
of 2009, when I got a call
from a provincial government
official, saying that the
province had decided they were
going to relinquish regulatory
control to the federal
government entirely, and I, my
understanding was that that
decision was not one that the
federal government necessarily
knew was coming.

ON MR. TAYLOR

PAN TO MS. MORTON

> MS. MORTON
> One of the complexities
> that jumped out at me is the
> province moved away from
> regulating this industry as
> soon as the Fraser sockeye
> crashed in 2009.

PAN BACK TO MR. TAYLOR

> MR. TAYLOR
> I'm going to, as Mr. Leadem did
> before me, ask the panellists
> to recognize that we're all
> time-limited and if you could
> respect the need to have some
> level of conciseness, please,
> in your questions, in your
> answers. Um, Ms. Morton, I'm
> going to read something, and
> I think you'll recognize the
> words that I read. "I am very
> interested in ensuring that
> aquaculture is properly managed
> and regulated and have a real
> concern that the Government
> of BC is acting outside its
> legal jurisdiction in its
> regulation of ocean aquaculture
> and the Government of Canada,
> as represented by Fisheries,
> has withdrawn from a proper
> regulatory role." Now, do you
> recognize those words?

PAN TO MS. MORTON

                    MS. MORTON
          No, sorry, I don't.

                    MR. TAYLOR (O.S.)
          Okay. If I said that's
          paragraph 12 of your affidavit
          in support of your litigation,
          does that refresh your memory?

                    MS. MORTON
          That would help, yes.

ON MR. TAYLOR

                    MR. TAYLOR
          And while you may not agree
          with the approach chosen, do
          you agree that an objective
          sound management and regulation
          of aquaculture . . .

PAN TO MS. MORTON

                    MR. TAYLOR (O.S.)
          . . . is a key component
          in attaining sustainable
          fisheries?

                    MS. MORTON
          It's key. I believe that the
          split mandate that DFO has is
          going to make that impossible.

BACK TO MR. TAYLOR

                    MR. TAYLOR
          I want to ask you about
          American University. That's
          where you got your degree,
          isn't it?

BACK TO MS. MORTON

                    MS. MORTON
          Yes, it is.

                    MR. TAYLOR (O.S.)
          And that's in Washington, DC?

                    MS. MORTON
          That's correct.

                    MR. TAYLOR (O.S.)
          And you obtained your degree in
          1977, did you?

                    MS. MORTON
          Yes, that's correct.

                    MR. TAYLOR (O.S.)
          That's a private university?

                    MS. MORTON
          Yes, it is.

                    MR. TAYLOR (O.S.)
          And is it known as famous for
          political activism?

                    MS. MORTON
          I don't know.

PAN BACK TO TAYLOR

                    MR. TAYLOR
          All right. What do you know its
          reputation to be?

                    MS. MORTON (O.S.)
          It was close to where my mother
          was living, and so that's where
          I began to take classes.

PAN TO MS. MORTON

                    MS. MORTON
          I hope we're going to get back
          to the sockeye here at some
          point.

QUICK PAN BACK TO MR. TAYLOR

                    MR. TAYLOR
          Is it what's referred to as a
          liberal arts college?

                    MS. MORTON (O.S.)
          I don't know.

                    MR. TAYLOR
          Okay.

The spectators grumble their disapproval from
the gallery.

PAN TO COMMISSIONER COHEN

                    COMMISSIONER COHEN
          Ladies and gentlemen, again,
          if I could ask you to
          respectfully honour the process
          here. We welcome the public's
          participation, and certainly
          we welcome your being in the
          public gallery, but if you
          would allow counsel to do their
          work, I would be very grateful.
          Thank you.

OVER TO MR. TAYLOR

                    MR. TAYLOR
          And that university has a
          college of arts and science,
          doesn't it, and did at the time
          you were there?

PAN TO MS. MORTON

                    MS. MORTON
          If you tell me it does, it did.

                    MR. TAYLOR (O.S.)
          Well, that's the college you
          were in, isn't it?

                    MS. MORTON
          Yes, but, you know, I was
          just taking my courses, going
          through it, and don't have a
                    (MORE)

                    MS. MORTON (CONT'D)
          recollection of exactly what
          that university was and all
          degrees and scope.

BACK TO MR. TAYLOR

                    MR. TAYLOR
          Did you get a bachelor's of
          arts and science?

PAN TO MS. MORTON

                    MS. MORTON
          I got a bachelor of science. I
          graduated magna cum laude.

                    MR. TAYLOR (O.S.)
          All right.

PAN TO MR. TAYLOR

                    MR. TAYLOR
          Will you agree with me that you
          are an advocate against open
          net fish farms?

BACK TO MS. MORTON

                    MS. MORTON
          I am an advocate for wild
          salmon.

                    MR. TAYLOR (O.S.)
          Okay.

Alexandra Morton walking with Chief Beau Dick and his family down Vancouver Island to Victoria to break a copper, a deeply traditional symbolic gesture against the government of British Columbia for its relationship with Indigenous people. (Photo credit: Anissa Reed)

PAN TO MR. TAYLOR

> MR. TAYLOR
> And is the corollary of that,
> that you're an advocate against
> open net pens?

> MS. MORTON (O.S.)
> As a corollary, yes. Because of
> the damage I see, I have become
> against net pen farms.

> MR. TAYLOR
> And you collaborate with
> other like-minded people in
> campaigning against active, or
> against open net pens, do you?

PAN TO MS. MORTON

> MS. MORTON
> I've collaborated with a wide
> range of people.

BACK TO MR. TAYLOR

> MR. TAYLOR
> All right. Do you have a blog?

> MS. MORTON (O.S.)
> Yes, I do.

> MR. TAYLOR
> And it's under the name
> Alexandra Morton?

                    MS. MORTON (O.S.)
          Yes, it is.

                    MR. TAYLOR
          Do you have control over the
          content?

                    MS. MORTON (O.S.)
          Yes, I do.

PAN TO MS. MORTON

                    MR. TAYLOR (O.S.)
          Do you know what's on it at any
          given time?

                    MS. MORTON
          Yeah, I'd have to refresh my
          memory to look at it, but yes.

                    MR. TAYLOR (O.S.)
          All right.

PAN TO MR. TAYLOR

                    MR. TAYLOR
          You, personally, put material
          on the blog, do you?

                    MS. MORTON (O.S.)
          Yes, that's correct.

                    MR. TAYLOR
          And over quite a long period
          of time you've been putting
          material on your blog to
          do with this Commission of
          Inquiry, haven't you?

                    MS. MORTON (O.S.)
          Yes, I have, because it's a
          public inquiry . . .

PAN TO MS. MORTON

                    MS. MORTON
          . . . that relates to wild
          salmon to which people are
          interested.

PAN TO MR. TAYLOR

                    MR. TAYLOR
          And you've put up material
          that, ah, is your account of
          the evidence given from day to
          day? Is that right?

PAN TO MS. MORTON

                    MS. MORTON
          Yes, that's correct.

AND RIGHT BACK TO MR. TAYLOR

This cartoon was posted by Alexandra Morton on her blog on August 08, 2011. It became part of the cross-examination of Morton and was seen as disparaging to other scientists appearing at the Inquiry. It was, perhaps, the lightest moment of the Inquiry when Morton was asked what is meant by "Liar, liar, pants on fire." (Credit: Anissa Reed)

                    MR. TAYLOR
          And your material on the
          blog includes commentary on
          witnesses?

                    MS. MORTON (O.S.)
          Yes, that's correct.

                    MR. TAYLOR
          And some of the commentary is
          quite disparaging?

            PAN TO MS. MORTON MS. MORTON
          Yes. It's been a disparaging
          experience.

DOCUMENT — Alexandra Morton's blog post of
the day in question.

                    MR. TAYLOR (O.S.)
          Mr. Lunn, are you able to pull
          up the September 8th blog?

                    MR. LUNN (O.S.)
          Yes.

                    MR. TAYLOR (O.S.)
          Do you recognize this as your
          blog, Ms. Morton?

                    MS. MORTON (O.S.)
          Yes, I do.

ON MR. TAYLOR

MR. TAYLOR
Is that a blog you posted
late last night or early this
morning?

MS. MORTON (O.S.)
Late last night, yes.

MR. TAYLOR
And you posted that blog
after you went under cross-
examination in these
proceedings?

MS. MORTON (O.S.)
Yes, that's correct.

PAN TO MS. MORTON

MR. TAYLOR (O.S.)
And you —

MS. MORTON
— I realize now that that was a
mistake.

PAN TO MR. TAYLOR

MR. TAYLOR
It's my understanding that
Mr. Martland or Ms. Grant
specifically told you not to
discuss your evidence with
anyone.

PAN TO MS. MORTON

                MS. MORTON
     Yeah, so it was the "discuss"
     that I made the mistake on,
     the back-and-forth. But since
     there's people sitting in the
     audience able to hear this,
     because it was live-streamed, I
     did not realize that there was
     that boundary.

BACK TO MR. TAYLOR

                MR. TAYLOR
     In this regard, though, you
     emailed Mr. Backman last
     evening, didn't you?

PAN TO MS. MORTON

                MS. MORTON
     Yes, because he made an
     interesting observation that
     I hadn't heard before, that
     the Harrison sockeye had been
     found going north, and I just
     had never heard that before,
     so I was just curious what his
     reference was for that.

BACK TO MR. TAYLOR

                    MR. TAYLOR
You'll agree with me that that
email you've just described
to Mr. Backman is a specific
reference to evidence in this
proceeding yesterday?

PAN TO MS. MORTON

                    MS. MORTON
Yes.

               MR. TAYLOR (O.S.)
After you'd been warned not
to discuss your evidence with
anyone, or any evidence with
anyone?

                    MS. MORTON
I asked him a question about a
reference that he had made. It
was not his opinion. And I do
apologize to the courts if I've
made a mistake here.

               MR. TAYLOR (O.S.)
We're kind of dancing on —

                    MS. MORTON
— But if I —

               MR. TAYLOR (O.S.)
— the head of a pin, aren't we? —

                    MS. MORTON
— was doing this in a, if I
knew I was breaking the rules,
I certainly would not have been
public with it. I would have
done something privately, but I
didn't realize I was breaking
the rules.

PAN TO MR. TAYLOR

                    MR. TAYLOR
Let's go to your blog of
September 8th, and I want
to specifically address the
bottom of page 1 and over onto
page 2. Now, Ms. Morton, that
paragraph: "I am really glad
Cohen will take a look at what
I pulled together." I take it
you mean the Commissioner in
these proceedings?

PAN TO MS. MORTON

                    MS. MORTON
Yes, that's correct. I was very
glad to hear that he was going
to read the report —

                    MR. TAYLOR (O.S.)
— Right.

What is happening to the Fraser sockeye?

Aquaculture Coalition
Based largely on documents submitted to the Cohen Commission
August 14, 2011

This document, *What is happening to the Fraser sockeye?*, was compiled by Alexandra
Morton and was highly contested by the Province of BC, the Canadian Government and the
BC Salmon Farmers Association when it was presented as a potential exhibit at the Cohen
Inquiry. It was eventually allowed to stand as an official exhibit. (Credit: Alexandra Morton)

                    MS. MORTON
          — because it's based on 500,000
          documents that were provided
          to this Commission, and I take
          my role seriously to offer all
          the information that he will
          need to make that decision
          as to whether aquaculture is
          impacting the Fraser sockeye.

OVER TO MR. TAYLOR

                    MR. TAYLOR
          Okay. As do many other people
          take their role seriously.
          You'll agree with that, will
          you?

                    MS. MORTON (O.S.)
          I, yes, I will agree with that.

                    MR. TAYLOR
          There's many people doing a lot
          of good and hard work in this
          Commission. Do you agree?

                    MS. MORTON (O.S.)
          There's many people, yes.

PAN TO MS. MORTON

                    MR. TAYLOR (O.S.)
          And the 500,000 documents you
          just referred to are mostly
          from the federal government,
          aren't they?

                    MS. MORTON
        Yes. Lots of provincial
        documents as well.

BACK TO MR. TAYLOR

                    MR. TAYLOR
        For your part, you produced
        about a couple of hundred
        documents, haven't you?

                    MS. MORTON (O.S.)
        That's right, because I'm
        not an expert on the Fraser
        sockeye.

                    MR. TAYLOR
        And yet you've got an awful lot
        of documents, don't you?

PAN TO MS. MORTON

                    MS. MORTON
        I have an awful lot of what
        type of documents?

                    MR. TAYLOR (O.S.)
        To do with Fraser sockeye and
        aquaculture.

                    MS. MORTON
        I do now, yes. I have —

                    MR. TAYLOR (O.S.)
        — No, no, never mind what you
        got through this Commission.

PAN TO MR. TAYLOR

                    MR. TAYLOR
          Apart from this Commission,
          you, yourself, and the
          Raincoast Research Society have
          a lot of documents on Fraser
          sockeye and aquaculture, don't
          you,

PAN TO MS. MORTON

                    MR. TAYLOR (O.S.)
          . . . from your work you've
          done?

                    MS. MORTON
          From the think tanks I've
          gone to and visiting the First
          Nations throughout the Fraser
          Valley last fall, I have lots
          of documents. And then on
          aquaculture, of course, I've
          got my own research and also an
          archive of scientific papers.
          I'm not really sure what you're
          getting at.

PAN TO MR. TAYLOR AND BACK

                    MS. MORTON
          I suspect this line of
          questioning is to prevent me
          from talking about what was
          actually in those documents.

BACK TO MR. TAYLOR

                    MR. TAYLOR
          Well, please don't try to
          speculate or worry about
          where I'm going with my
          questions. I'll just ask if
          you could answer them. Now,
          still with your documents, do
          you understand, when I say
          "documents," I'm including
          electronic material such as
          emails?

PAN TO MS. MORTON

                    MS. MORTON
          Okay.

                    MR. TAYLOR (O.S.)
          Do you understand that?

                    MS. MORTON
          I understand that.

                    MR. TAYLOR (O.S.)
          And you're quite a prolific
          emailer to do with Fraser
          sockeye and aquaculture, aren't
          you?

                    MS. MORTON
          Yes, I am, because working
          through government processes
          and working through science
                    (MORE)

                    MS. MORTON (CONT'D)
          didn't work, so it has pushed
          me to another phase where I
          feel that the public need to
          hear from me directly.

                    MR. TAYLOR (O.S.)
          All right.

PAN TO MR. TAYLOR

                    MR. TAYLOR
          And you've produced virtually
          none of your documents to this
          Commission. Is that right?

BACK TO MS. MORTON

                    MS. MORTON
          Yeah. They're in my blogs,
          though.

BACK TO MR. TAYLOR

                    MR. TAYLOR
          All right. Now, you say, in
          your blog, which we'll return
          to now, in that sentence
          beginning "I'm really glad,"
          "The report is not my work,
          it's a compilation of what DFO
          has been saying." The report
          you're referring to is the
          document that was spoken of
          yesterday that Mr. McDade
                    (MORE)

                    MR. TAYLOR (CONT'D)
          tried to get into evidence?
          That's the report that's being
          referred to here in the report,
          in the blog, isn't it?

PAN TO MS. MORTON

                    MS. MORTON
          Yes, correct.

BACK TO MR. TAYLOR

                    MR. TAYLOR
          And then you go on and you
          state some things and state
          them as fact, and, um, you'll
          see, at the bottom of the page
          on the screen, it says: "A
          DFO scientist tasked to find
          out, find out why millions of
          sockeye are dying just before
          spawning found evidence that
          a virus associated with cancer
          is killing them — fact." That
          statement is wrong, isn't it?

PAN TO MS. MORTON

                    MS. MORTON
          It's a fact that she has found
          evidence.

WHIP BACK TO MR. TAYLOR

> MR. TAYLOR
> You heard Dr. Miller's
> testimony last week, didn't
> you?

WHIP BACK TO MS. MORTON

> MS. MORTON
> Yes, I did.

> MR. TAYLOR (O.S.)
> And you heard Dr. Garver's
> evidence?

> MS. MORTON
> Yes, I did.

> MR. TAYLOR (O.S.)
> And both of them said, "This is
> a work in progress," and they
> are not making, they have not
> reached the conclusion that you
> put here?

> MS. MORTON
> I don't think "evidence"
> means a conclusion. It means
> evidence.

ON MR. TAYLOR

                    MR. TAYLOR
     And over the page your blog
     says: "The only known place a
     virus like this occurs is in
     the salmon farms on the dying
     sockeye's migration route —
     fact."

BACK TO MS. MORTON

                    MR. TAYLOR (O.S.)
     You know that to be wrong,
     don't you?

                    MS. MORTON
     No, I don't. I don't know
     anywhere else where marine
     anemia, salmon leukemia,
     plasmatoid leukemia, parvo, any
     of those things have ever been
     reported, other than the papers
     from Dr. Kent, Dr. Stephens,
     Dr. Ribble and others.

PAN TO MR. TAYLOR

                    MR. TAYLOR
     Well, you know that the work
     that Dr. Miller is doing is to
     do with a syndrome that she's
     identified, correct?

PAN TO MS. MORTON

MS. MORTON

You know, the word "syndrome"
related to marine, marine
anemia did not arise until
this Commission, until several
scientists were on the stand.
Back in the days when they
were simply writing about it,
Dr. Kent actually named it the
salmon leukemia virus.

MR. TAYLOR (O.S.)

I'm talking about Dr. Miller's
work.

BACK TO MR. TAYLOR

MR. TAYLOR

You know that it's called a
syndrome? She is the scientist
and that's what she's termed it
as, correct?

BACK TO MS. MORTON

MS. MORTON

No, I don't think she is
calling it a syndrome. In a lot
of her work, like Exhibit 613G,
she ponders salmon leukemia
virus —

MR. TAYLOR (O.S.)

— All right.

                    MS. MORTON
— and she points to it as
coming, as the only known
source was the salmon farms.

BACK TO MR. TAYLOR

                    MR. TAYLOR
I have your evidence, thank
you. The next line says:
"DFO's response? Cut off the
researcher's funding." You know
that to be wrong. That's not
what Dr. Miller said, is it?

PAN TO MS. MORTON

                    MS. MORTON
No, she did. She said —

                    MR. TAYLOR (O.S.)
— No —

                    MS. MORTON
— "I don't have any funding to
go further on sockeye."

                    MR. TAYLOR (O.S.)
And you know and you heard her
evidence . . .

BACK TO MR. TAYLOR

                    MR. TAYLOR
. . . that her staff are at
work, being paid. The problem's
been identified as a rules-
related problem, it's being
worked on and a fix has to be
found. You've heard all of that
evidence, didn't you?

PAN TO MS. MORTON

                    MS. MORTON
I heard her say she no longer
had funding to work on sockeye.

PAN TO MR. TAYLOR

                    MR. TAYLOR
And then you were blind to all
of the rest of the evidence
I've just said?

                MS. MORTON (O.S.)
I can only believe what she
said.

BACK TO MS. MORTON

                    MS. MORTON
Now, she is going to hopefully
go look for parvo in salmon
farms —

                MR. TAYLOR (O.S.)
— Would it be more —

                    MS. MORTON
          — after the aquaculture
          hearings are over, but what she
          said on the stand, and I'm sure
          we can find her testimony, is
          that she had no further funding
          to work on sockeye.

PAN TO MR. TAYLOR

                    MR. TAYLOR
          Would it be accurate to say
          you just don't pay attention to
          what you don't what to hear?

BACK TO MS. MORTON

                    MS. MORTON
          I don't think you can hear me.
          We should pull up her testimony
          right now and check that out.

PAN TO MR. TAYLOR

                    MR. TAYLOR
          Yeah, I'll ask the questions,
          please. Now, this blog —

The crowd grumbles again.

                    MS. MORTON (O.S.)
          — Is that going to stand in
          this court?

                    MR. TAYLOR
Mr. Commissioner, this blog was
done contrary to the rules that
are in place for this inquiry,
so I'm in your hands as to what
to do with it at this point. I
don't want to make an exhibit
that, which is a violation of
the rules of this inquiry. I
tend to think it should be an
exhibit for identification.
I've read in what I need to,
and leave it at that.

PAN TO COMMISSIONER COHEN

                COMMISSIONER COHEN
What's the next identification
number, Mr. Registrar?

            REGISTRAR GILES (O.S.)
GGG.

                COMMISSIONER COHEN
Thank you.

            REGISTRAR GILES (O.S.)
Triple G.

BACK TO MR. TAYLOR

                    MR. TAYLOR
          Mr. Lunn, may we have the blog
          that is August 31st, please?
          You recognize that as your blog
          of August 31st, Ms. Morton?

                    MS. MORTON (O.S.)
          Yes, I do.

DOCUMENT — Exhibit 1839, Alexandra Morton's
blog post "Cohen Inquiry Aquaculture Hearings
August 31."

                    MR. TAYLOR
          This, Ms. Morton, deals
          with the evidence that the
          veterinarians gave on August
          31st, doesn't it?

PAN TO MS. MORTON

                    MS. MORTON
          Yes, that's correct.

BACK TO MR. TAYLOR

                    MR. TAYLOR
          This is a cartoon of what
          appears to be the Commissioner
          speaking to those four
          witnesses. Is that what that
          is?

PAN TO MS. MORTON

                    MS. MORTON
         Yes, that's correct.

DOCUMENT — The cartoon drawing appears on
screen.

                    MR. TAYLOR (O.S.)
         And the cartoon is showing
         flames coming from the pants of
         the witnesses, correct?

                    MS. MORTON
         Yes.

The crowd in the gallery laughs.

BACK TO MR. TAYLOR

                    MR. TAYLOR
         So we've got Swerdfager,
         Sheppard, McKenzie and Marty,
         and the cartoon is showing
         them with pants on fire. And
         the word that the Commissioner
         says, in the cartoon, that is,
         "pants on fire." What does that
         mean?

PAN TO MS. MORTON

                    MS. MORTON
         Well, I'm going to leave that
         to you. It just —

                    MR. TAYLOR (O.S.)
          — Well, you're familiar with
          the saying —

                    MS. MORTON
          — My — the reason —

                    MR. TAYLOR (O.S.)
          — "Liar, liar, pants on fire"?

                    MS. MORTON
          The reason that I put this up
          is because Dr. Gary Marty is
          reporting symptoms of a disease
          that's of enormous significance
          to this Commission —

                    MR. TAYLOR (O.S.)
          — Okay, let me ask you this —

                    MS. MORTON
          — and yet Dr. Sheppard does not
          acknowledge that that disease
          exists.

                    MR. TAYLOR (O.S.)
          Yeah, that's all fine, we've
          heard that.

BACK TO MR. TAYLOR

                    MR. TAYLOR
          But let me ask you this: Do
          you agree with me that that
          cartoon is disparaging of those
          witnesses' evidence?

BACK TO MS. MORTON

                    MS. MORTON
        I felt it was a representation
        without saying the words.

                 MR. TAYLOR (O.S.)
        Are you saying they lied?

                    MS. MORTON
        How can you look at the
        symptoms of a disease, have
        somebody like Gary Marty report
        those symptoms —

                 MR. TAYLOR (O.S.)
        — My, my —

                    MS. MORTON
        — as being the clinical signs
        of marine anemia, which a DFO
        scientist thinks the majority
        of Fraser sockeye are being
        killed and weakened by, and
        the vets above him, Peter
        McKenzie of Mainstream, and
        Dr. Mark Sheppard, simply don't
        recognize that that disease
        exists? That is —

                 MR. TAYLOR (O.S.)
        — Ms. Morton —

                    MS. MORTON
        — it cannot stand.

BACK TO MR. TAYLOR

> MR. TAYLOR
> Do you agree that that cartoon
> is disparaging of those
> witnesses' evidence?

PAN TO MS. MORTON

> MS. MORTON
> No. I think that their jobs
> force them into that position
> and I feel sorry for them.

BACK TO MR. TAYLOR

> MR. TAYLOR
> Do you agree with me that it is
> against the code of conduct for
> a registered biologist to speak
> disparagingly of a colleague
> registered biologist?

PAN TO MS. MORTON

> MS. MORTON
> It is, yes.

> MR. TAYLOR (O.S.)
> And can we equally apply
> that, then, to you should
> not be disparaging of other
> professionals, such as
> veterinarians?

                    MS. MORTON
Mr. Taylor, my personal code of
conduct, is it when I see —

                    MR. TAYLOR (O.S.)
— No, I'm asking about the
biologists' code of conduct —

                    MS. MORTON
— an ecosystem being destroyed,
I will use what tools I can
that are fair and legal to try
to represent that truth.

WHIP BACK AND FORTH

                    MR. TAYLOR (O.S.)
All right. Thank you. I have
your evidence.

                    MS. MORTON
And if a cartoon was the only
way I could do it, that's what
I was going to do.

                                    DISSOLVE TO:

BACK TO MR. TAYLOR

                    MR. TAYLOR
Do you believe that sea lice
are generated by farms, then
transmitted to wild, and then
kill wild stocks in sufficient
                    (MORE)

                    MR. TAYLOR (CONT'D)
          numbers to have a measurable
          and significant negative effect
          on wild sockeye populations?

BACK TO MS. MORTON

                    MS. MORTON
          Yes.

                    MR. TAYLOR (O.S.)
          So you disagree with the likes
          of Dr. Korman, Noakes, Connors,
          Jones, Beamish, Hargreaves,
          Johnson?

                    MS. MORTON
          You're asking me if that is
          possible and I said, "Yes." If
          there are enough lice on those
          fish —

                    MR. TAYLOR (O.S.)
          — I see. —

                    MS. MORTON
          — it would definitely kill them.

                    MR. TAYLOR (O.S.)
          Thank you.

PAN TO MR. TAYLOR

                    MR. TAYLOR
So that's your belief. That's
your perspective, is it? Is
that right?

                    MS. MORTON (O.S.)
Well, I mean, you put me on
this panel as a layman,

PAN TO MS. MORTON

                    MS. MORTON
. . . but you don't recognize
that I've done over 20 papers
on sea lice.

                    MR. TAYLOR (O.S.)
Well —

                    MS. MORTON
— So I've done a lot of work
where I've actually viewed
the impact of the lice on pink
and chum salmon, but also on
sockeye. But I haven't done
the experimental work with the
sockeye, holding them as I did
with the pink and chum.

                    MR. TAYLOR (O.S.)
All right.

BACK TO MR. TAYLOR

                    MR. TAYLOR
          Thank you. I believe your
          participation on this panel
          is important, but just to be
          clear, I didn't put you on the
          panel, the Commission counsel,
          of course, did.

                                    DISSOLVE TO:

                    MR. TAYLOR
          Finally,

ON MS. MORTON

                    MR. TAYLOR (O.S.)
          . . . if we could go to Exhibit
          DDD, page 59 or 60.

BACK TO MR. TAYLOR

                    MR. TAYLOR
          At the bottom of that page you
          say: "Canada has no mechanism
          to react to the threat of
          exotic viruses that are
          traveling in farmed salmon eggs
          worldwide." You know full well
          that there's quite a rigorous
          egg importation protocol and
          regime in place in British
          Columbia, don't you?

                    MS. MORTON (O.S)
          No, sir, there is not.

                    MR. TAYLOR
All right.

                    MS. MORTON (O.S)
The fish health certificate
does not have infectious salmon
anemia on it.

                    MR. TAYLOR
All right. Thank you. I have
your evidence on that.

ON MS. MORTON

                    MR. TAYLOR (O.S.)
Then you say,

BACK TO MR. TAYLOR

                    MR. TAYLOR
. . . "DFO policy is to promote
salmon farms. They are being
pressured . . .

ON MS. MORTON

                    MR. TAYLOR (O.S.)
. . . by the salmon farming
corporations to do so and
field staff seem unable to
communicate accurately about
salmon farm impacts." That's
your interpretation of DFO
policy I take it, is it?

                    MS. MORTON
          That's my personal experience.

                    MR. TAYLOR (O.S.)
          All right.

PAN BACK TO MR. TAYLOR

                    MR. TAYLOR
          But that doesn't accord with
          the evidence in this inquiry
          from people such as Mr.
          Thompson or Mr. Swerdfager,
          does it?

PAN BACK TO MS. MORTON

                    MS. MORTON
          Well, if you would allow my
          60-page document to go in as
          evidence, there is evidence
          there. I was also a reviewer
          for Dr. Beamish's paper with
          the *ICES Journal of Marine
          Science*. I also viewed Dr.
          Jones's laboratory experiment
          on juvenile pink salmon sea
          lice, so I actually have
          enormous experience and I'm
          sorry that we weren't able to
          talk to that, to speak to that.

MR. TAYLOR (O.S.)
And at this point my time is up
as well. So I will sit down.

FADE TO:

MR. KELLIHER
Panellists, my name is . . .

**LOWER THIRD: Steven Kelliher, Counsel,
Aboriginal Aquaculture Association.**

MR. KELLIHER
. . . Steven Kelliher, and I
am counsel for the Aboriginal
Aquaculture Association.

ON MS. MORTON

MR. KELLIHER (O.S.)
Now, bearing in mind this is a
perspectives panel, I'm going
to ask . . .

BACK TO MR. KELLIHER

MR. KELLIHER
. . . you a question, each of
you, that I asked a series
of scientists that have been
on the stand, testified over
the last few days. Uh, their
names were Dr., Drs. Korman,
O'Connors, Dill, Noakes, Jones,
(MORE)

MR. KELLIHER (CONT'D)
Saksida and Orr. Each of them,
with I think the largest
qualification being Dr. Orr's,
gave a positive answer to the
proposition, the question of
whether open pen fish farming
could on the west coast of
British Columbia coexist with
thriving wild stocks.

ON MS. PARKER

MS. PARKER
Yes, I think that fish farms
can coexist with wild stocks.
I think that's partly because
of the precautionary framework
towards management that we have
in place. I think it's because
of the adaptive management.

ON MS. MORTON

MS. PARKER (O.S.)
I think it is because we have
the science . . .

BACK TO MS. PARKER

MS. PARKER
. . . and the ability to make
good decisions. We have risk-
based management. And, and,
(MORE)

                    MS. PARKER (CONT'D)
uh, with all that in place, we
cancontinue to have coastal
employment. And I think one of
the values of salmon farming
is it's not just minimum wage
jobs in coastal communities.
It is highly skilled technical
positions.

                    MR. KELLIHER (O.S.)
Ms. Stewart.

PAN TO MS. STEWART

                    MS. STEWART
I think many of the answers
that you got were very
qualified, and I heard Mike
Price and others saying that
it, it's entirely theoretical.
They could potentially coexist
if, for example, there was one
open net cage farm in an entire
region like the Broughton
Archipelago.

                    MR. KELLIHER (O.S.)
I didn't refer to Mr. Price.

                    MS. STEWART
Well, okay. But I'm just saying
that I think that all of the
various answers were often
                    (MORE)

                    MS. STEWART (CONT'D)
          qualified. Um, and I would
          also suggest that the
          question would have to engage
          around the current levels
          of production or potential
          increases in production.
          But I'd just add, as well,
          that I understand fully the
          importance of employment in
          those communities. I know Percy
          Starr, as well, and spent many,
          many hours and days and time
          in the Kitasoo community, and
          Mr. Starr was very clear that
          this was a choice that they
          were largely forced into due to
          the collapse of the wild salmon
          stocks and they couldn't go
          fishing.

ON MR. KELLIHER

                    MS. STEWART (O.S.)
          And honestly, I believe that
          if DFO would support the wild
          stocks . . .

BACK TO MS. STEWART

                    MS. STEWART
          . . . to the extent that
          they support the aquaculture
          industry, perhaps communities
                    (MORE)

                    MS. STEWART (CONT'D)
        wouldn't be faced with those
        choices. It's their right to
        make that decision.

                    MR. KELLIHER (O.S.)
        Yes or no, Ms. Stewart?

                    MS. STEWART
        What's the question? Can they
        coexist?

PAN BACK TO MR. KELLIHER

                    MR. KELLIHER
        It's not wholly an opportunity
        to make speeches, the idea is
        to eventually get around to
        answering the question.

                    MS. STEWART (O.S.)
        Yeah. It may be possible —

                    MR. KELLIHER
        — Mr. Backman? —

                    MS. STEWART (O.S.)
        — said at the beginning, it may
        be possible for them —

                    MR. KELLIHER
        — Mr. Backman, can I ask you —

PAN TO MS. STEWART

                    MS. STEWART
          — to coexist if there is
          a serious limit on the
          production, and at current
          levels, I don't believe so.

PAN TO MR. BACKMAN

                    MR. KELLIHER (O.S.)
          All right.

THEN TO MR. KELLIHER

                    MR. KELLIHER
          It's possible if it were limited
          and best practices. Is that
          right?

                    MS. STEWART (O.S.)
          Seriously limited.

                    MR. KELLIHER
          Seriously limited. Mr. Backman?

OVER TO MR. BACKMAN

                    MR. BACKMAN
          I think that the Report 5
          results have shown us that
          currently the information that
          was shared, that aquaculture
          is coexisting with the wild
          fish without demonstrated
          significant risk of disease, I
          think that the answer to
                    (MORE)

                    MR. BACKMAN (CONT'D)
          your question is yes, the
          possibility is there, is there
          now and it remains and going
          into the future, with, keep
          up the standards that we have
          today, and we keep the actions
          in place to look at and reduce
          any liabilities that come up.

PAN TO MS. MORTON

                    MR. KELLIHER (O.S.)
          All right. Ms. Morton.

                    MS. MORTON
          There are no First Nations on
          this coast that want to see the
          Fraser sockeye wiped out.

                    MR. KELLIHER (O.S.)
          There's nobody in this room
          that does either, Ms. Morton.

                    MS. MORTON
          We're not talking about Marine
          Harvest and Grieg employees at
          this point, but I believe the
          answer to your question is no.
          This industry cannot survive
          biologically. There are viruses
          attacking this industry around
          the world, and what are those
          towns going to do when what
          happened to Chile happens again
          here?

PAN TO MR. KELLIHER

          MR. KELLIHER
    All right. So you tell the
    First Nations, such as Kitasoo,
    to pull their nets out of
    the water and close down the
    processing plants. Is that
    right?

BACK TO MS. MORTON

She takes a deep breath, speaking
passionately.

          MS. MORTON
    If I had a choice between the
    wild salmon and the ability
    to bring them back, and an
    industry that brings salmon
    from the Atlantic and feeds
    them on fish from Chile, in a
    small port town like Kitasoo
    and uses them as an example
    that all other First Nations
    are supposed to swallow, with
    the scientists that have been
    up here before, you have preyed
    on their respect for First
    Nations. Out of respect for
    First Nations, they acquiesced
    to you. You're a very skilled
    lawyer. But what about the
    people of the Broughton? What
          (MORE)

                    MS. MORTON (CONT'D)
          about the people that are in
          the audience right now who have
          said no to the industry and are
          being run over as if they don't
          count. What about them?

BACK TO MR. KELLIHER

                    MR. KELLIHER
          Can you, can you explain this
          to me, Ms. Morton? The names
          that I read out earlier are
          well-respected scientists with
          a very significant history
          and body of knowledge in this
          sphere. All of them carry PhDs.

ON MS. MORTON

                    MR. KELLIHER (O.S.)
          All of them say that the . .

BACK TO MR. KELLIHER

                    MR. KELLIHER
          . . . wild stocks can coexist
          within water nets. You are the
          only one that says no. Why is
          that?

PAN TO MS. MORTON

                    MS. MORTON
          That's because I don't work
          for a university. I don't work
          for the Government of Canada.
          I don't work for the Province
          of BC. I don't work for a
          First Nations community. I am
          completely independent. I might
          be the only independent —

BACK TO MR. KELLIHER

                    MR. KELLIHER
          — You are pure, are you?
          You're the only one that isn't
          corrupted by business, by
          government, by a university. Is
          that correct?

OVER TO MS. MORTON

                    MS. MORTON
          Perhaps.

                    MR. KELLIHER
          All right.

OVER BLACK

**SUPERSCRIPT: Examination by Morton
Counsel . . .**

FADE IN:

INT. COURTROOM — DAY

The gallery is full. Commissioner Cohen enters.

                    MR. MARTLAND (O.S.)
          Mr. Commissioner,

ON MR. MARTLAND

                    MR. MARTLAND
          . . . we move, next, to counsel
          for the Aquaculture Coalition
          with 90 minutes.

ON MR. MCDADE

                    MR. MCDADE
          Thank you, Mr. Commissioner. My
          name is Greg McDade.

LOWER THIRD: Gregory McDade, QC, Counsel, Aquaculture Coalition/Alexandra Morton.

                    MR. MCDADE
          I'm counsel for the Aquaculture
          Coalition, so Dr. Morton is one
          of my clients. Um, let me start
          with you, Dr. Morton, and get
                    (MORE)

                    MR. MCDADE (CONT'D)
          a sense of how involved in
          this matter you've been. I
          understand you spent a fair bit
          of time looking through the
          Ringtail database?

BACK TO MS. MORTON

                    MS. MORTON
          Yes, that's correct.

                                        FADE TO:

**POP-UP — Ringtail database: Secure online
database of documents for legal proceedings**

ON MR. MCDADE

                    MR. MCDADE
          And as a result of all of your
          research in the database and
          your extensive investigations, I
          gather you've got a perspective
          on the issues before the
          Commission, in terms of what's
          happened to the ah, to the
          sockeye since 1992 and, in fact,
          what happened to the sockeye in
          2009, and you've prepared your
          evidence in written form.

DOCUMENT — The front page of Exhibit BBB.

> MR. MCDADE (O.S.)
> Can we have Exhibit BBB up on
> the screen?

PUSH IN — The title "What is happening to the
Fraser sockeye?" fills the screen.

> MR. MCDADE (O.S.)
> So this is a document,

BACK TO MR. MCDADE

> MR. MCDADE
> . . . I gather, that really
> encapsulates the evidence you
> want to give today?

PAN TO MS. MORTON

> MS. MORTON
> That's correct.

> MR. MCDADE (O.S.)
> And you wrote this document
> yourself?

> MS. MORTON
> I wrote it myself.

> MR. MCDADE (O.S.)
> Yes. And you adopt it as your
> evidence?

> MS. MORTON
> Yes, I do.

                    MR. MCDADE (O.S.)
          Can I have that marked as the
          next exhibit, please?

PAN TO MR. TAYLOR

                    MR. TAYLOR
          I'm objecting. I'll let others —

A number of speakers try to talk at the same
time.

                    MR. MCDADE (O.S.)
          I see my friends from the
          Salmon Farmers and the Province
          on their feet as well.

PAN TO MR. BLAIR

                    MR. BLAIR
          I'm objecting as well. For the
          record, Alan Blair, for the BC
          Salmon Farmers Association.

PAN TO MR. PROWSE

                    MR. PROWSE
          I'm objecting as well, My Lord,
          or Mr. Commissioner.

OVER TO MR. TAYLOR

                    MR. TAYLOR
     Well, I'll go first, but I
     think we should start with
     knowing what the Commission's
     position is.

Mr. Taylor looks toward Mr. Martland.

PAN TO MR. MARTLAND

                    MR. MARTLAND
     It's easy for me to do this,
     Mr. Commissioner. I would like
     to hear what the basis for
     the objection is. We've taken
     a broad approach, generally
     speaking. We can anticipate
     what some of the concerns may
     be. I would like to learn what
     the, ah, objection taken to
     this document is. It is a case
     where notice has been given,
     the witness is here.

PAN BACK TO MR. TAYLOR

                    MR. TAYLOR
     Well, I'll go first, and we can
     probably proceed in the order
     of our participant number. As
     Mr. McDade has elicited in
     evidence, this is a document
     that the witness prepared for
     this particular inquiry. It's
                    (MORE)

                    MR. TAYLOR (CONT'D)
said to be her evidence. She's
here to give her evidence
viva voce, not to tender a
written document. And before
I forget, I want to point out
that if this document were
to go in, Mr. McDade would
then leave it to all of us
to have to cross-examine on
a lengthy document not put in
through viva voce, the facts
not put in through viva voce
evidence, and we don't have
time to do that. So it would
leave evidence essentially, un-
not cross-examined, and that
wouldn't be right or fair, in
my submission. But there's
more substantive reasons for it
as well. This document, which
we've reviewed, is Ms. Morton's
own account or review of
documentation that she's looked
at. And she then puts her
interpretation on the documents
and her understanding and her
views and so forth. Again, to
the extent they're her views,
this is the Perspectives Panel
and that can be elicited viva
voce. To the extent that it's
Ms. Morton's interpretation of
documents, that's for you,
                    (MORE)

                    MR. TAYLOR (CONT'D)
          Mr. Commissioner, in the final
          analysis, and for counsel along
          the way in submissions to do
          it, but it's not for witnesses,
          in my submission.

ON MR. BLAIR

                    MR. BLAIR
          Mr. Commissioner, Alan Blair,
          for the BC Salmon Farmers
          Association. I've read the
          document and clearly it purports
          to be a quasi-expert report,
          and I note that this panel
          is specifically before this
          Commission not as qualified
          experts but for their unique
          and individual perspectives
          on the matter. The document's
          full of hearsay and speculation.
          There are science conclusions
          that she draws which are far
          beyond her expertise. But what
          is of most concern to me and
          to this process, I think, if I
          may, and we don't need to pull
          it up on the screen now, but
          we can, because we filed it,
          Ms. Morton's, um, curriculum
          vitae starts with, registered
          professional biologist since
          1988. And the code of conduct
                    (MORE)

                    MR. BLAIR (CONT'D)
          lists what conduct a registered
          professional biologist is
          entitled to, to, how they are
          to conduct themselves. And
          among other matters, and we
          can pull it up paragraph by
          paragraph, but a professional
          biologist is to be objective
          and honest in all matters,
          reports, testimony. Objective
          is certainly not what this
          document is.

ON MR. PROWSE

                    MR. PROWSE
          Mr. Commissioner, I adopt
          my friend from Canada's
          objections. Fundamentally, this
          is a, a document which I think
          we can anticipate will largely
          form the basis of written
          submissions and perhaps oral
          submissions at the end of the
          day. It's not a document that I
          submit qualifies as evidence in
          this hearing. And secondly, the
          document is certainly verging
          on purporting to provide expert
          opinion evidence on matters
          particularly of disease which
          are well outside of this
          witness's realm of expertise.
          So I adopt the objections of my
          friend from Canada.

PAN TO MR. LEADEM

                    MR. LEADEM
          Mr. Commissioner, Leadem,
          initial T., appearing as
          counsel for the Conservation
          Coalition.

LOWER THIRD: Tim Leadem, Counsel,
Conservation Coalition.

                    MR. LEADEM
          I'll be very brief. I think
          this is not a question of
          admissibility so much as it is
          a question of probative value,
          and I think that once you
          determine that it's admissible,
          and I, I could hear no ground
          upon which it's inadmissible,
          other than fairness, and we're
          all operating under the same
          time constraints here, I wish
          that I could have had a lot
          more time to cross-examine many
          of the witnesses that preceded
          these witnesses to the forum.
          So it really goes to weight, it
          doesn't go to admissibility,
          so I'd ask you to allow it to
          be admitted. You yourself can
          judge its probative value at
          the end of the day.

People in the gallery clap.

ON MR. MARTLAND

                    MR. MARTLAND
          Mr. Commissioner, I'm going
          to, through the court, ask
          members of the gallery simply
          to do their best, I appreciate
          it may be exciting, or it may
          not be, but I'll ask folks,
          nonetheless, to please abstain
          from making noise during these
          proceedings. Ms. Gaertner had a
          further point.

ON MS. GAERTNER

                    MS. GAERTNER
          Mr. Commissioner, I've
          canvassed with all the counsel
          of First Nations so that you
          only have to hear from one of
          us,

LOWER THIRD: Brenda Gaertner, Counsel, First Nations Coalition.

                    MS. GAERTNER
          . . . and we actually, having
          not heard what Mr. Leadem
          was going to say, adopt his
          position, and submit that this
          is a matter of weight. As it
          relates to the issue of
                    (MORE)

                MS. GAERTNER (CONT'D)
        fairness, I can only emphasize
        that we have all been operating
        with a significant challenge
        of trying to pick and choose
        what we have time to make
        submissions on and what do we
        have time to actually cross-
        examine on. And so, what's good
        for the goose is good for the
        gander on that one.

PAN TO MR. MARTLAND

                MR. MARTLAND
        Mr. Commissioner, from our
        point of view, this, the
        question of admissibility
        is one that falls to your
        discretion to be determined. In
        my respectful submission . . .

PAN TO COMMISSIONER COHEN

                MR. MARTLAND (O.S.)
        . . . given the well-
        established practice in public
        inquiries which, generally
        speaking, take a much broader
        approach to questions of
        receivability or admissibility,
        and not even evidence of
        information,

BACK TO MR. MARTLAND

                    MR. MARTLAND
          . . . there are examples
          of public inquiries for
          commissioners who have boarded
          planes and sat in meeting rooms
          to receive relevant information.
          So there's a broad process
          and a broad approach. This
          Commission has not taken that
          approach. This Commission, in
          a number of respects, is trial-
          like. The rules do permit
          evidence and information to be
          brought forward in a flexible
          way. We've certainly seen, and
          I think we've increased the
          pace of, in recent days of the
          number of reports, whether
          they're reports written for the
          Commission or things published
          in academic journals that are
          put in, in a very quick fashion.
          Mr. Taylor's point about being
          able to meaningfully cross-
          examine on the entirety of a
          document is true. Generally
          speaking, though, that's been
          a disadvantage to folks like
          Mr. McDade and Mr. Leadem in
          not being able, for example, to
          cross-examine Dr. Saksida on six
          or eight journal articles that
          she has written which are now
                      (MORE)

MR. MARTLAND (CONT'D)
in evidence. In our respectful
submission, to echo what Mr.
Leadem had to say, the concerns
identified about there being
views or interpretations,
concerns about underlying
documents or underlying facts
or previous testimony, concerns
about this being submissions as
opposed to argument, all speak
to the question of weight. They
are not decisive in terms of
the question of admissibility.

ON COMMISSIONER COHEN

COMMISSIONER COHEN
Mr. McDade?

PAN TO MR. MCDADE

MR. MCDADE
I simply agree, it's just a
matter of weight, and point
out that we have not had the
luxury of being able to call
witnesses, witnesses, and we
do not have the luxury of being
able to choose time. This is
the only way we get this story
and this perspective before
the court — or before the
Commission.

PAN BACK TO COMMISSIONER COHEN

COMMISSIONER COHEN
Yes, thank you, counsel, for
your very helpful submissions.
Um, I think you heard Mr.
Martland say yesterday, we have
a number of exhibits, now, that
have, as you know, been marked
for identification,

**POP-UP — Marked for identification: Documents
that must become full exhibits before they
are used as evidence**

COMMISSIONER COHEN
. . . and they fall into
different categories in terms
of the areas in which they
have been entered and, to
some extent, in a few cases
there have been submissions
directly with respect to
the admissibility of those
exhibits. In other cases,
it was simply marked for
identification and left for
a later time to deal with.
My ruling is this: Ah, Mr.
Martland, and I think he is
attempting to work with you in
arriving at an omnibus position
with respect to exhibits
that have been marked for
identification, which many of
you here want to have marked
(MORE)

COMMISSIONER COHEN (CONT'D)
as exhibits, may or may not
be able to reach an accord
with all of you on that matter
pertaining to the exhibits you
wish to have marked. I frankly
doubt that this particular
document will reach an
accord, simply because of the
positions you have taken here
this afternoon. So I am going
to leave it, for the moment,
marked for identification
purposes. I have not read
this document. I don't know
what's in it. I only know what
Mr. Taylor and Mr. Blair have
alluded to and, of course,
the rest of you. Mr. Leadem,
of course, raises a, a point
which I, in general, agree
with, but as I haven't seen
this document, I haven't heard
the testimony of Dr. Morton, I
don't know, frankly, how to put
the emphasis or non-emphasis
on the positions they've taken.
So this document is going
to remain for identification
purposes. Mr. McDade has a full
opportunity to elicit evidence
from Dr. Morton. All of you
will have an opportunity to
cross-examine her in due
(MORE)

COMMISSIONER COHEN (CONT'D)
course, and if you don't reach
an omnibus position, I will
issue a separate ruling with
respect to the admissibility
of this particular document. I
don't feel it would be fair,
frankly, at this point to
simply enter it without giving
consideration to all of your
remarks, and once I have an
opportunity to consider this
document as well. Thank you.

BACK TO MR. MCDADE

MR. MCDADE
Well, in that circumstance,
then, um, Mr. Commissioner, I
think it's important that I
spend a little more time on the
document than I would otherwise
have done. We may have to go
through it page by page.

DISSOLVE TO:

TIGHTER ON MR. MCDADE

MR. MCDADE
And the next section of your
paper deals with pre-spawn
mortality, and you cite a
number of pieces of information
(MORE)

> MR. MCDADE (CONT'D)
> that you've learned from the
> Ringtail database. Can you
> describe that for us?

OVER TO MS. MORTON

> MS. MORTON
> Yes. I really didn't know that
> much about pre-spawn mortality,
> but in reading through Ringtail
> I became highly educated on it.
> Um, if you look at the third
> paragraph down: "Since 1995, an
> average of 58% and up to 95% of
> the late run sockeye have died
> of pre-spawn mortality."

ON MR. MCDADE — He's listening intently.

> MS. MORTON (O.S.)
> And they're wondering if it's
> a freshwater parasite, called
> *Parvicapsula*, which, when
> I wrote to Brian Riddell a
> couple of years ago, that was
> his response back to me, they
> thought it was *Parvicapsula*
> because the fish were heavily
> infected with it, and so that,
> in opening them up, that's what
> drew their attention first.

BACK TO MR. MCDADE

MR. MCDADE
Can you describe the chart that
you've got there?

DOCUMENT — Exhibit BBB, page 7 graph.

MS. MORTON (O.S.)
Yes. So this was also presented
in 2009, and it's astonishing.
So these orange bubbles show
the amount below average
returns for each of those runs,
and you see they're broken down
into the . . .

BACK TO MS. MORTON

MS. MORTON
. . . names of the runs, but
also clumped as to whether they
were Early, Summer or Late.

PANNING DOWN THE GRAPH — The highlight stops
on the Harrison River run.

MS. MORTON (O.S.)
And astonishingly, one of
them is incomplete, running
completely contrary to the
others, and that's the
Harrison.

BACK TO MS. MORTON

                    MS. MORTON
So it does make a biologist
wonder, what, what is different
about those fish?

ON MR. MCDADE

                    MS. MORTON (O.S.)
The Harrison have two very
different life histories
strategies.

ON MS. MORTON

                    MS. MORTON
One is they leave the river
when they're very small, like
the pink and chum, so that
means they, right now, what's
happening is the Fraser sockeye
adults are passing salmon farms
and they're coming in from
all over the open ocean and
they're going into the river
and they're going straight
into the nursery areas of the
Fraser sockeye that are raising
in there as smolts. But the
Harrison are gone. They left
already in May and June. So
they don't get that exposure.
Plus, they're not going by the
salmon farms. So, for me, this
was an astonishing graph.

During the Cohen Inquiry, it was reported that the Harrison River sockeye population had higher than normal recruitment for a number of years. At the time, I was still searching for a suitable ending to my film *The Pristine Coast*, so I took a trip out to the Harrison River to see for myself. I was approached there by fisherman François Perreault, who told me there were thousands of salmon dying and floating down the river. This photo was taken at the headwaters of the Harrison River where it flows out of Harrison Lake, September 10, 2011. (Photo credit: Scott Renyard)

Looking toward the east bank of the Harrison River headwaters, where many of the dead salmon got caught in the back eddy, September 10, 2011. It was clear to me that the Harrison River salmon runs may have been the only good news story prior to the Cohen Inquiry. Ironically, that all changed while the Inquiry hearings were being held. See Appendix B for more information and photographs. (Photo credit: Scott Renyard)

Looking downstream toward the east bank of the Harrison River
headwaters, where hundreds of dead sockeye salmon are tangled in freshwater
aquatic plants, September 10, 2011. (Photo credit: Scott Renyard)

Many of the dead sockeye were bloated. Their flesh would split and bulge out, which
is not what you usually see when a salmon dies after spawning. (Photo credit: Scott Renyard)

ON MR. MCDADE

                    MR. MCDADE
          The next topic, yes, is, is
          related to Dr. Kristi Miller's
          MRS study.

POP-UP — MRS: Mortality related signature

PAN TO MS. MORTON

                    MS. MORTON
          Yes.

                    MR. MCDADE (O.S.)
          You took an interest in that
          through your database research?

                    MS. MORTON
          Yes. So, first of all, I was
          shocked to run across her work
          in Ringtail, because I sat in
          the 2009 Simon Fraser University
          think tank, and we were supposed
          to figure out what had happened
          to Fraser sockeye and report
          back to the public, and we were
          never told about this work. So
          basically what happened was so
          many sockeye were dying in the
          Fraser River, the DFO realized
          that before they could open a
          fishery they would have to figure
          out how many were going to die
                    (MORE)

                    MS. MORTON (CONT'D)
          in the river, and so they
          tasked their genomic profiler,
          Kristi Miller, to try to figure
          that out, and they, by the
          sounds of it and reading their
          work, they thought they were
          going to, she was going to find
          that they ran out of steam,
          that they hadn't fed well
          enough. But she stumbled on
          the truth, uncomfortable truth,
          that there was a pattern which
          she said looked like a virus.

ON COMMISSIONER COHEN — He sits with his
hands folded, listening intently to Morton's
testimony.

                    MS. MORTON
          She's like, "Okay, well, it
          looks like a retrovirus."

**POP-UP — Retrovirus: Viruses that use host
DNA to replicate**

                    MS. MORTON
          There's only two retroviruses in
          salmon. And, "Oh, look, one of
          them occurs in salmon farms that
          went into the Fraser sockeye
          migration in 1992," so —

                                        DISSOLVE TO:

ON MR. MCDADE

               MR. MCDADE
   — Does it matter what we call it?

PAN TO MS. MORTON

               MS. MORTON
   It doesn't matter what we call
   it. And so it was interesting,
   on the stand, to hear these
   men to say they don't, they
   actually don't know what it is.
   They never went the final stage
   to visualize it. They, they
   never were able to actually
   figure out if it was a virus
   or not. They're calling it a
   syndrome.

BACK TO MR. MCDADE

               MR. MCDADE
   "These gentlemen never figured
   it out." I think you were
   referring to the plasmacytoid
   leukemia or the marine anemia
   back in the late '90s or —

PAN TO MS. MORTON

               MS. MORTON
   — That's right, so —

            MR. MCDADE (O.S.)
— early — the early '90s —

            MS. MORTON
— Dr. Kent pioneered it. He
was director of the Pacific
Biological Station. During a
portion of that he actually
named it plasmatoid leukemia.
Dr. Craig Stephens did his
PhD thesis, so as another,
Dr. Ribble, and they very
helpfully, in a 1997 paper,
gave us a diagnostic, because
they said it's a difficult
thing to understand and to
diagnose, so they said, if
there's interstitial cell
hyperplasia of the caudal
kidney, you've probably got it.

PAN TO MR. MCDADE

            MR. MCDADE
So whatever that disease was
that they were talking about
in the 1990s — sorry, I see
another objection.

PAN TO MR. PROWSE

MR. PROWSE
Yes, Mr. Commissioner. I simply
want to state on the record
that for the reasons that I
objected to the written report,
I think much of what we're
hearing is a dialogue between
counsel and the witness about
evidence which ought, in my
submission, to be, at the
end of the day, a submission
between counsel for the
participant and the Commission,
and I don't think this is
evidence. And asking what her
"perspective" is, quote unquote,
I don't think really advances
the matter. So I object to the
line of questioning on that
basis.

BACK TO MR. MCDADE

MR. MCDADE
I'll take that as a statement
for the record and proceed,
Mr. Commissioner. The, um, in
your research, did you see a
connection between plasmacytoid
leukemia and BKD?

PAN TO MS. MORTON

                    MS. MORTON
There was a lot of reference to
plasmatoid leukemia causing a
swelling in the kidney, and a
lot of confusion in the early
days. They, they expressed
this. And I guess the reason
that Mr. McDade and I are
trying to do this is because
I spent so much time reading
all this. You really need to
know it too, and I don't want
you to have to go through the
2,000 hours, but when you look
at what they did in the Fraser
River, they tested those Cultus
sockeye dozens of times for
BKD . . .

POP-UP — BKD: Bacterial kidney disease

                    MS. MORTON
. . . and the tests came up
negative, negative, negative.
They were losing 100 percent
three years in a row, according
to Mike Lapointe, of pre-spawn
mortality. They thought it was
BKD, but it wasn't. So when
I heard that salmon leukemia
looks like BKD, it did make me
wonder, and then, with Miller's
work on top of it, it does make
a person wonder if that's what
they were dying of.

PAN TO MR. MCDADE

              MR. MCDADE
And when you went to the fish
health database, what did you
find?

WHIP PAN TO MS. MORTON

              MS. MORTON
I, well, the early, the early
records from the salmon farming
were scattered and very hard
to interpret, but when I came
across BCP002864 written by
Dr. Gary Marty, there I had
something I could really just
look at. And he, I found that
the symptoms that Drs. Kent
and Stephens and Ribble had
said are the diagnostic simple
for marine anemia, they're
being diagnosed regularly. And
in the abbreviations tab, Dr.
Marty was saying, "These can
be associated with a clinical
diagnosis of marine anemia."
But then I heard Dr. Sheppard
say he doesn't believe in
marine anemia and he's not
going to diagnose it on a farm
level. So it's not going to
appear on the records. But Dr.
Marty is seeing something, and
              (MORE)

                    MS. MORTON (CONT'D)
we do need to figure out what
that is, and he is seeing it in
chinook farms, in particular,
in higher severity, but he's
also seeing it in the Atlantic
farms as a level 1 severity.

                    MR. MCDADE
Describe the point you're
making here, Dr. Morton?

A whip pan turns into a graph.

GRAPH — Recruits per spawner over time.

                    MS. MORTON (O.S.)
Yes. This was presented at the
SFU think tank that I attended
in 2009, and it shows that the
productivity, which is the
number of spawners returning
from each female sockeye,
has been dropping since
approximately 1992 in quite a
precipitous manner.

ON MS. MORTON

                    MR. MCDADE (O.S.)
The Pacifics are the chinook
salmon?

                    MS. MORTON
That's right.

                    MR. MCDADE (O.S.)
          That were present in the, ah,
          in the Discovery Islands?

                    MS. MORTON
          Yes. So —

                    MR. MCDADE (O.S.)
          — What's the significance of
          this finding for you?

                    MS. MORTON
          Well, the next thing I became
          curious about, of course, is
          when you're looking at the
          pattern of the Fraser sockeye,
          you do want to wonder what
          happened in 2010.

**POP-UP — 2010: Fraser River sockeye returned
to normal recruitment levels**

                    MS. MORTON
          And so when I saw Miller's
          work and I saw what Kent and
          Stephens said about this
          disease, that it spreads from
          chinook to sockeye, they
          actually tested that, that
          it was lethal to sockeye, I
          wondered, well, how many
          chinook farms are there on the
          Fraser sockeye migration routes
                      (MORE)

> MS. MORTON (CONT'D)
> along eastern Vancouver Island?
> And so I went to the database
> that Josh Korman . . .

GRAPH — Number of chinook salmon farms by year in Discovery Islands.

> MS. MORTON (O.S.)
> . . . organized, and he lists
> whether they're Atlantic or
> chinook, so I just looked
> at the chinook farms. And
> interestingly enough, after
> June of 2007, there have been
> no . . .

BACK TO MS. MORTON

> MS. MORTON
> . . . chinook farms on the
> Fraser sockeye migration route.
> And so the fish that came back
> in 2010 went by no chinook
> farms and were not exposed to
> these numbers and the severity
> of these symptoms and they came
> back. Now, this is for somebody
> else to figure out, but this
> is the pattern that I'm able to
> read from these databases and
> from the information that's in
> Ringtail.

ON MR. MCDADE

                    MR. MCDADE
          Perhaps this is, since this
          is an appropriate time. Well,
          first of all, can I mark that,
          now, as a full exhibit now, Mr.
          Commissioner?

PAN TO MR. TAYLOR

                    MR. TAYLOR
          We've been through this before.
          This is how it got to be
          exhibit for identification, and
          it's part of this, what's being
          called, omnibus approach to see
          what we do with this and many
          other exhibits.

OVER TO MR. MCDADE

                    MR. MCDADE
          I don't think that's true, Mr.
          Commissioner. It was because
          Dr. Morton hadn't been on the
          stand yet, that's why it was
          for identification.

BACK TO MR. TAYLOR

                    MR. TAYLOR
          I mean, still, you can do many
          things with numbers, and we
          heard some of that just now
          from Ms. Morton —

The spectators laugh.

BACK TO MR. MCDADE

                    MR. MCDADE
          There have been many similar
          exhibits marked by previous
          witnesses. I don't understand
          this at all.

BACK TO MR. TAYLOR

                    MR. TAYLOR
          Well, I'm still speaking, until
          the gallery interrupted.

More noise from the gallery.

                    MAN IN GALLERY
          Sorry.

                    WOMAN IN GALLERY
          Public inquiry.

TO COMMISSIONER COHEN

                    COMMISSIONER COHEN
          Ladies and gentlemen, I think
          Commission counsel have asked,
          respectfully, that you honour
          the process here, and I would
          be very grateful if you could
          withhold any comments while
          you're in the public gallery.
                    (MORE)

COMMISSIONER COHEN (CONT'D)
It would be very helpful for
all of us. Thank you very much.
Mr. Taylor?

BACK TO MR. TAYLOR

MR. TAYLOR
Thank you. Ms. Morton spoke
to, you could look at the
information she had and put it
together this way or that, she
was speaking to that a few
moments ago. This is properly,
in my view, something that
should remain an exhibit for
identification. We can deal
with it later. It's not simply
a matter of arithmetic. We're
bordering on expert evidence
at the moment, and it should
remain for identification.

BACK TO COMMISSIONER COHEN

COMMISSIONER COHEN
Mr. McDade, I really don't want
to cut into your time. I think
these kinds of documents may,
in the end, all be marked, but
I think we're going to run into
this difficulty, and I would
respectfully suggest that we
move on. We will deal with
(MORE)

COMMISSIONER COHEN (CONT'D)
these. If I have to make
separate rulings on these, I
will. Hopefully counsel will
work out an understanding about
marking these documents, but in
the meantime, I'd like you to
move on.

BACK TO MR. MCDADE

MR. MCDADE
All right. So you were going,
Dr. Morton, you were going to
take us to the next page?

GRAPH — ISAV-like lesions and marine anemia
symptoms diagnosed in farm salmon off eastern
Vancouver Island.

MS. MORTON (O.S.)
So this is the exotic virus. It
spikes the quarter before the
infectious salmon anemia-like
lesions.

BACK TO MS. MORTON

MS. MORTON
And so I've pondered this a
long time, had it up on my wall,
just considering it, and the
marine anemia syndrome is noted
by many scientists to be an
(MORE)

                    MS. MORTON (CONT'D)
          immune-suppressing situation
          for the fish. It takes a co-
          factor to actually kill the
          fish. So BKD,

POP-UP — BKD: Bacterial kidney disease

                    MS. MORTON
          . . . loma,

POP-UP — Loma: Loma gill disease

                    MS. MORTON
          . . . different parasites, IHN,

**POP-UP — IHN: Infectious hematopoietic
necrosis virus**

                    MS. MORTON
          . . . all of these things
          attack a fish more easily
          if it's weakened with marine
          anemia, and —

                    MR. PROWSE (O.S.)
          — Mr. Commissioner,

PAN TO MR. PROWSE

                    MR. PROWSE
          . . . I rise to object again.
          The witness is now getting into
          questions of interpretation of
                    (MORE)

                    MR. PROWSE (CONT'D)
          disease and that's outside her
          field of expertise, and I think
          that's a significant objection
          that must be made.

ON COMMISSIONER COHEN

                    COMMISSIONER COHEN
          Mr. McDade?

PAN TO MR. MCDADE

                    MR. MCDADE
          Again, Mr. Commissioner, we
          sought to qualify Dr. Morton
          as an expert and Commission
          counsel said that she'd be
          called at the Perspectives
          Panel and able to give her
          perspectives on these
          matters. And I don't know why
          counsel for the Province is
          so determined to keep this
          information from you, but it
          should be allowed.

PAN BACK TO COMMISSIONER COHEN

                    COMMISSIONER COHEN
          I think so long as it's
          made clear that this is a
          perspective and not an opinion,
          Mr. McDade, it is fine.

                    MR. MCDADE (O.S.)
          Yes.

                    COMMISSIONER COHEN
          I did start to collect the
          impression that we were moving
          off the perspective kind of
          evidence. Thank you.

PAN TO MR. MCDADE

                    MR. MCDADE
          Yes, I'm simply trying, Mr.
          Commissioner, to put some of
          these questions in front of
          you. The conclusions to be
          drawn from them are yours.

                    MS. MORTON (O.S.)
          Yeah, the only thing I would
          want . . .

PAN TO MS. MORTON

                    MS. MORTON
          . . . you to take this, take
          from this, Mr. Commissioner,
          is that Dr. Miller needs a
          chance to look at what this
          is. Somebody, somebody more
          experienced with disease, who's
          known to speak freely, needs to
          look at this.

ON COMMISSIONER COHEN

He contemplates Ms. Morton's statement.

                                        FADE OUT:

OVER BLACK

SUPERSCRIPT: In the end, Alexandra's report was allowed to stand as an exhibit.

                                        FADE TO:

SUPERSCRIPT: In 2015, Alexandra took salmon farmers to court because they were putting diseased Atlantic smolts in their open net pens.

                                        FADE TO:

SUPERSCRIPT: She won the case and the judge gave fish farmers 4 months to remove the diseased fish from the water.

                                        FADE TO:

SUPERSCRIPT: The salmon farmers have appealed.

ROLL END CREDITS

EXT. VANCOUVER ART GALLERY PROTEST — DAY

Alexandra Morton is under a tent on a stage. It's pouring rain and she's speaking to a crowd of protesters before the start of the Cohen Inquiry hearings.

MS. MORTON

Hi! I thank you all for standing
there in the rain. I know that's
not comfortable. But we are the
wild salmon people.
            (cheers)
Every one of us is worth 20
of the people that come in the
sunny days. So, there's just a
couple of things I want to say.
One is, it's our right to have
wild salmon. We don't even need
to find a reason, but of course
there are reasons. They're good
for our economy, they're good
for our environment, they feed
our soul. I have started the
wildsalmonpeople.ca so that we
can continue to be united, form
a voting block and get more
politicians in there like Fin
Donnelly. People who understand
us and work for the people,
because this is not just about
salmon. It's about the democracy
that we're losing. It's about
our ability to survive. It's
about our communities, and
I want every one of you to
know that you hold the power
to change this. We are not
livestock. We can think, we can
vote and we can do as we did
today. Powerfully, peacefully
            (MORE)

                    MS. MORTON (CONT'D)
make ourselves visible and
make ourselves clearly
understood. So, join me at
wildsalmonpeople.ca. I'm going
to interview every politician
in British Columbia, federal
and provincial, and I'm going
to ask them how they plan to
let us survive because as the
wild salmon go, so we go. We
are the top carnivore of this
province. We need everything
underneath of us. I have just
travelled a thousand kilometres
up the Fraser watershed. The
last bit I did with Margo
French, careening around about
on some dirt roads above the
Alaskan border, and every
single spawning grounds that
these salmon go to, there
are people standing guard
because they understand. It's
phenomenal to be part of you.
And thank you so much to all of
you paddlers that went down the
river with me. I love all of
you. Thank you so, so much.

                    THE END

Appendix A

This document is the memorandum of understanding that was signed between the Province of British Columbia and the Government of Canada on September 8, 1988. It transferred most of the responsibility for the growing aquaculture industry from the federal government to the Province of British Columbia. Overlapping jurisdictions in the marine waters of Canada was the impetus driving the two levels of senior levels of government to establish a better way to regulate the new industry. This MOU was overturned by the Morton decision when it was found to be unconstitutional. Alexandra Morton felt that in spite of the arrangement, the aquaculture industry was not being regulated properly and wild salmon populations were being damaged by the industry.

CANADA/BRITISH COLUMBIA MEMORANDUM OF UNDERSTANDING
ON AQUACULTURE DEVELOPMENT

THIS AGREEMENT made this   6   day of Sept 1988

BETWEEN THE GOVERNMENT OF CANADA (hereinafter referred to as "Canada" represented by the Minister of Fisheries and Oceans)

OF THE FIRST PART

AND THE GOVERNMENT OF BRITISH COLUMBIA (hereinafter referred to as "British Columbia") represented by the Minister of Agriculture and Fisheries

OF THE SECOND PART

WHEREAS Canada and British Columbia wish to establish a mutual agreement to advance the orderly growth and development of the aquaculture industry in British Columbia;

AND WHEREAS both Canada and British Columbia have substantial interests in the prudent development of an economically sound aquaculture sector and the facilitation of investment therein;

AND WHEREAS both Canada and British Columbia are interested in identifying and clarifying their respective roles in advancement of the aquaculture sector;

THEREFORE, without prejudice to their respective Constitutional powers, the parties hereby agree:

1. Definitions

1.1 "aquaculture" means the commercial culture or husbandry of aquatic plants or animals by the private sector, and excludes ocean ranching, spawn-on-kelp, bait-holding, and contracted Salmonid Enhancement Program activities;

1.2 "aquaculture facility" means lands or waters where aquaculture is conducted, including the buildings, improvements or fixtures situated thereon, and associated personal property such as ships, barges, tanks, cages or structures, but excludes fish processing facilities;

1.3 "plants" and "animals" includes adult forms, seed, eggs, larvae, young or juvenile stages or any parts thereof;

1.4 "extension program" means the provision and promotion of technical, marketing, production and business management information, procedures and technologies to and for the benefit of the aquaculture industry; and

1.5 "technology transfer" means transmittal and adoption of scientific and production information, procedures and technologies to and for the benefit of the aquaculture industry.

2. Scope

2.1 This Agreement extends to the aquatic plants and animals currently, or which may be, husbanded or cultured in British Columbia.

2.2. Nothing in the Agreement affects or extends to the terms and conditions of valid Provincial tenure or to applications, policy, procedures, guidelines or legal duties and obligations concerning that tenure. "Provincial Tenure" means the right to occupy Provincial Crown Lands, however conferred, and includes permission conferred by a licence issued under authority of the Land Act.

3. Research and Development

3.1 British Columbia and Canada shall facilitate research, development and technology transfer as it relates to aquaculture, in a co-operative manner to ensure that programs are complementary, cost effective and meet the needs of industry. The Management Committee (Schedule A) shall be the forum for co-ordination, co-operation and the identification of priorities.

177

3.2   Canada shall undertake research and development related to the aquaculture industry in British Columbia including conducting research at aquaculture facilities where appropriate. Canada will conduct basic and applied mission-oriented research related to the aquaculture industry. Canada shall communicate research and development results to the aquaculture industry and British Columbia, and shall encourage rapid communication of key results to those parties.

3.3   British Columbia will facilitate and encourage the development, provision and delivery of extension programs to the aquaculture industry, and Canada shall co-operate with British Columbia in the provision of those services, and in facilitating technology transfer to the industry.

3.4   British Columbia and Canada agree to promote and encourage the aquaculture industry, universities and other agencies to undertake, or focus, activities related to the development, acquisition, application and dissemination of knowledge or technology to maximize the benefit of such activities to the aquaculture industry.

4. Education and Training

4.1   British Columbia will facilitate and encourage the development, provision and delivery of educational and on-the-job programs that will provide academic, technical and safety training for the aquaculture industry.

4.2   Canada will encourage the training of graduate students, post-doctoral fellows, technical or industry personnel in its research facilities and in other institutions.

5. Administration of Aquaculture

5.1   Licensing and Regulation: Provincial

5.1.1   British Columbia may issue licences to carry out aquaculture operations in the Province of British Columbia.

5.1.2   The Ministry of Agriculture and Fisheries will act as Provincial lead agency in dealing with Canada. However, any ministry of the Provincial Government may be responsible for exercising its powers and regulations herein referred to, and may deal directly with Canada.

5.1.3   British Columbia, in establishing regulations and policies for the aquaculture industry, may address, among others, the following concerns:

        (a) development and management of the aquaculture industry in the Province;

        (b) establishment of categories of aquaculture licences and the terms and conditions of each;

        (c) exemption of persons, classes of persons, types of aquaculture or activities from provincial regulations;

(d) fees or royalties in regard to licensing aquaculture operations;

(e) prescription of forms or applications for aquaculture licences and for approvals to transport and transfer aquatic plants and animals within British Columbia;

(f) size, spacing, density and location of aquaculture facilities and the use, content and enforcement of site-development plans;

(g) number of aquaculture licences that may be held by one person;

(h) marking and identification of aquaculture sites and structures for purposes other than the protection of navigation;

(i) reporting requirements, records and documents, and fees in respect thereof;

(j) performance standards for aquaculture facilities;

(k) qualifications or financial standards for aquaculture facilities;

(l) protection of the confidentiality of information required from licencees and applicants;

(m) methods of handling, buying, selling, holding and possession, offering or advertising for sale or maintaining the quality of aquatic plants or animals within the province;

(n) approved methods of harvesting in an aquaculture facility and prohibitions of such harvesting without the consent of the licencee;

(o) standards relative to the design, layout, construction materials and equipment of aquaculture facilities.

5.1.4 In undertaking the above, British Columbia recognizes the need for orderly and responsible growth and development of aquaculture and agrees to consult with Canada to develop criteria and standards that recognize the possible impacts of aquaculture and to minimize adverse affects of aquaculture on fish health, fish stocks, fish habitat and fishing activities.

5.1.5 All aquaculture licence applications in British Columbia shall be referred to Canada for comment prior to establishing the conditions of licences.

5.2. Regulation: Federal

5.2.1 The Department of Fisheries and Oceans will act as lead federal agency for aquaculture in British Columbia. However, any federal department of the Government of Canada may be responsible for exercising its powers and regulations herein referred to, and may deal directly with British Columbia in that regard.

- 4 -

5.2.2   Leases issued by Canada over Federal properties
         shall continue to be administered by Canada.

5.2.3.   The Fish Health Protection Regulations and related
         instruments under the _Fisheries Act_ shall apply
         to all stocks in aquaculture facilities.

5.2.4   Canada may enact regulations for conservation
         and protection of wildstocks and fish habitat
         with respect to aquaculture.

5.2.5   Canada will continue to facilitate the provision
         of approvals under the _Navigable Waters Protection_
         Act of aquaculture facilities to ensure that
         other waterway users have equitable access and
         that marine navigational rights of passage are
         respected.

5.2.6   For federally regulated species, Canada is
         responsible for issuing permits for collecting
         wild broodstock for aquaculture including eggs,
         milt, spawn, larvae, juveniles and adults.

5.2.7   Canada undertakes to carry out or cause to be
         carried out sanitary shellfish water quality
         surveys, paralytic shellfish poison surveys,
         and product and other certification programs
         in accordance with all shellfish export
         requirements.

5.3 Co-ordination

5.3.1   Upon execution of this Agreement, Canada shall
         advise holders of existing licences, issued by
         Canada, to apply for a licence issued by British
         Columbia. In replacing the licences formerly
         issued by Canada, British Columbia will undertake
         to honour the general purposes and conditions
         of those licences.

5.3.2   Canada and British Columbia will develop mutually
         acceptable aquaculture referral processes that
         consider fish health, fish habitat and fish har-
         vesting concerns.

5.3.3   British Columbia and Canada will consult in exer-
         cising existing and in establishing new regulations
         and policies for the aquaculture industry and
         may address, among others, the following concerns:

         a) the introduction into the province and transfer
             and transport within the province between
             aquaculture facilities and to processing
             plants within the province, of aquatic plants
             and animals including conducting environmental
             assessment relative to such activities;

         b) isolation and quarantining of aquatic plants
             and animals, and where diseased or infested
             with harmful lifeforms, disposal or destruction
             of such plants and animals, and disinfection
             or disposition of equipment related to such
             plants and animals;

         c) standards relative to construction and opera-
             tion of aquaculture facilities, and methods
             of handling, storage and use of chemicals,
             fertilizers, vaccines, feeds and other sub-
             stances used in the conduct of aquaculture;
             and

d) in conjunction with the aquaculture industry, development of product quality standards.

5.3.4 Canada agrees to provide, where possible, both general and site-specific information necessary to identify critical fish habitats, stocks and related matters to enable British Columbia to develop policies and programs that minimize the possibility of adverse impacts from aquaculture.

5.3.5 Canada and British Columbia will consult on the development of policy regarding international and interprovincial importations of aquatic plants or animals or intra-provincial movements which would expose a watershed or marine area, all or part of which, is included in British Columbia, to the introduction of exotic species, diseases and pests.

5.3.6 Canada and British Columbia will, through the Federal-Provincial Transplant Committee, review, approve or reject applications for the introduction, transport and transfer of aquatic plants and animals into and within British Columbia.

5.3.7 Canada and British Columbia will establish mechanisms for on-going dialogue with the aquaculture industry, in a form and with terms of reference as agreed to by the Management Committee.

5.4 Dispute Resolution

5.4.1 In the event of a dispute between Canada and British Columbia over a substantive question affecting matters referred to in this agreement it shall be referred to the Management Committee. Where the Management Committee is unable to resolve the dispute, it shall be referred to the Deputy Minister of Fisheries and Oceans for Canada and the Deputy Minister of Agriculture and Fisheries for British Columbia who shall diligently attempt to resolve the dispute as quickly as possible and and in accordance with the intent of this Agreement.

5.4.2 Where a court of competent jurisdiction finds a particular regulation to be ultra vires the powers of Canada or British Columbia and neither government intends to appeal the decision or the appeal process has been exhausted, the government that has jurisdiction for the matter shall consider forthwith the passing of substantially similar regulations to replace the ones declared ultra vires by the Court.

5.4.3 Notwithstanding anything in this Agreement, Canada or British Columbia may take measures deemed necessary to protect matters within its jurisdiction.

5.5 Compliance and Inspection

5.5.1 Canada and British Columbia shall conduct periodic inspections of aquaculture facilities to determine compliance with their respective Acts, regulations and guidelines and will provide the other with results relevant to their mandate of those inspections. Nothing in this Agreement shall affect the duties of the parties with regard to fish plant inspections.

5.5.2   Canada will consult with British Columbia concerning appointment, as agents for fish health protection purposes, of qualified persons recommended by British Columbia. British Columbia shall have a role in the detection, prevention, control and eradication of fish diseases in British Columbia.

5.5.3   Canada and British Columbia in accordance with their respective environmental and fish habitat protection mandates, shall from time to time, monitor waste accumulations from aquaculture operations and the effects thereof, and share this information with each other.

5.5.4   (a) Canada will exercise its responsibilities to monitor aquaculture products destined for human consumption for antibiotic residues, toxic materials and other additives or contaminants likely to pose a hazard to human health;

         (b) Canada will exercise its responsibilities to specify quality or grade standards for interprovincial and international trade and inspect for compliance; and

         (c) British Columbia will exercise its responsibilities to license and inspect facilities buying, vending and processing aquaculture products for intraprovincial trade.

5.5.5   Nothing in the above shall affect the existing arrangements regarding the enforcement of the international agreement on molluscs, and Canada will continue to exercise control over exportation of molluscs.

5.5.6   Canada and British Columbia shall consult to establish effective procedures for inspection and enforcement.

6.   **Feed for Aquaculture**

6.1.   In developing and implementing policies to optimize the allocation, harvest and utilization of fish stocks and fish offal, Canada agrees to take into consideration fish feed requirements of the British Columbia aquaculture industry.

7.   **Egg Supply**

7.1   Canada and British Columbia will negotiate annually the quantity of salmon eggs to be made available to the aquaculture industry.

7.2   Within the constraints of salmon conservation and stock rebuilding, Canada will make available small quantities of genetic material from wild salmon stocks to facilitate aquaculture broodstock development.

8.   **Therapeutants and Vaccines**

8.1   Therapeutic drugs used in aquaculture shall be regulated by Health and Welfare Canada. Vaccines used in aquaculture shall be regulated by

Agriculture Canada. Canada agrees to facilitate research, development and testing in these areas.

9. Statistics

9.1   Aquaculture products statistics shall be collected as specified in the existing Canada-British Columbia Letter of Agreement on Co-operative Collection and Analysis of Processed Fish Statistics.

9.2   British Columbia shall collect annually, in a mutually agreed form, data from aquaculture facilities relevant to production, distribution and sales. It shall provide the data to Canada.

9.3   Canada shall collect annually, in a mutually agreed form, data on aquaculture imports and exports.

9.4   Canada shall compile provincial statistics on aquaculture and publish them annually in a national report together with its own statistics.

9.5   To facilitate the exchange and supply of cultivated stock, Canada shall develop and maintain a National Registry of important aquaculture stocks. Canada shall make available to British Columbia information concerning those stocks, whether or not indigenous to British Columbia, including information about each stock's performance characteristics, ancestry and related facts.

9.6   Canada shall maintain Pacific and National Registries of fish diseases and a data centre for the documentation and dissemination of information pertaining to fish diseases in Canada.

10. Management and Implementation

10.1   There shall be a Management Committee whose structure and functions are as set out in Schedule A.

10.2   The Director General, Pacific Region, Department of Fisheries and Oceans, shall represent Canada for the implementation of this Agreement on behalf of Canada, and shall be Co-Chairman of the Management Committee.

10.3   The Deputy Minister, Ministry of Agriculture and Fisheries, or designate, shall represent British Columbia for the implementation of this Agreement on behalf of British Columbia and shall be Co-Chairman of the Management Committee.

10.4   Implementation of this Agreement will be co-ordinated with other Canada-British Columbia agreements administered by the Department of Fisheries and Oceans.

10.5   The parties agree that they shall use their best efforts to achieve expeditious alteration of legislation or administrative policies that may impede the implementation of this Agreement.

10.6   The Management Committee will meet semi-annually to review implementation of this Agreement and will consult as necessary to ensure its effective operation.

10.7 The parties will use their best efforts to ensure implementation of the intent of this Agreement in fostering growth and development of aquaculture in British Columbia.

11. National Co-ordination

11.1 The parties will co-operate with other provinces, if possible, through national meetings or other arrangements, for the putting in place of plans and projects aimed at developing aquaculture and at promoting a co-ordinated and joint approach to the development of aquaculture and the marketing of its products.

12. Amendments to Agreement

12.1 This Agreement may be amended at any time on mutual accord.

12.2 A notice of proposal to amend the Agreement by one party shall be submitted in writing to the other party, which shall respond within three (3) months. Failure to respond within three (3) months shall be deemed to be rejection of the amendment proposal.

13. Coming into Force

13.1 This Agreement shall come into force on signing.

14. Termination

14.1 The present Agreement may be terminated on one year's written notice by either party.

14.2 Licences issued during this Agreement remain valid for one year, or until their date of expiry, whichever comes first, notwithstanding that the Agreement may be terminated during this period.

SIGNED IN THE PRESENCE OF: GOVERNMENT OF CANADA

"Original signed by the
Honourable Tom Siddon"

Witness _____   Minister of Fisheries and Oceans

GOVERNMENT OF BRITISH COLUMBIA

"Original signed by the
Honourable John Savage"

Witness _____   Minister of Agriculture and Fisheries

SCHEDULE A

MANAGEMENT COMMITTEE

The Management Committee shall be comprised of equal numbers
of federal and provincial members. It shall have a least 4
members. It shall meet not less than semi-annually.

The Management Committee shall

- function as a co-ordination and liaison mechanism to
  implement this Agreement;

- identify priorities, timing, sequence and funding for
  activities of joint interest;

- co-ordinate and consult with industry and other interested
  groups including non-government or international organizations;

- strike and co-ordinate subordinate committees or task
  groups as necessary to perform its duties;

- identify research priorities and encourage timely communication of key research results to the industry;

- develop terms of reference to establish and maintain
  a direct communications link with industry; and

- function in resolving disputes arising between Canada
  and British Columbia.

Appendix B

When I discovered Harrison River salmon runs were dying in large numbers in 2011 it was too late for me to make a submission to the Cohen Inquiry. I decided to call Alexandra Morton to talk about what I had discovered. Morton immediately made arrangements to come to Vancouver, and I took her and Anissa Reed to the Harrison River to show them the salmon die-off. Morton subsequently returned to the river and took samples from sockeye, coho, chinook, pink, and chum salmon that had died before spawning. The photographs below show some of the salmon found and sampled by Morton and her team.

After discovering that the Harrison River sockeye were not faring as well as reported, I reached out to Alexandra Morton, who brought a team to investigate. Here she is retrieving a coho salmon from the Harrison River on October 13, 2011. (Photo credit: Anissa Reed)

Alexandra Morton retrieving a jaundiced chinook
salmon, October 13, 2011. (Photo credit: Anissa Reed)

Not all fish floating down the Harrison River in 2011 were fresh from the ocean. But even sockeye that were close to spawning were not able to complete their life cycle. Dr. Kristi Miller discovered the cause of these deaths was a virus with linkages to leukemia and the salmon farming industry. (Photo credit: Anissa Reed)

Alexandra Morton and her team collected a number of salmon from the Harrison River which were then taken to a dock so the team could gather samples. In this photo, three species of salmon were found dead and floating in the river: sockeye, chinook and coho. During other sampling trips, chum and pink salmon were also affected by the die-off. This means that all five Pacific salmon species suffered losses in the Harrison River system in 2011. (Photo credit: Anissa Reed)

Alexandra Morton on a Harrison River beach, opening the bellies of the dead salmon. All died before spawning, a phenomenon known as pre-spawn mortality. These fish all died before completing their life cycle. (Photo credit: Anissa Reed)

The sockeye salmon marked #1 was one of the fish Alexandra Morton's team took samples from for testing, October 12, 2011. (Photo credit: Anissa Reed)

Alexandra Morton opens the belly of coho #1 to discover it was full of eggs. (Photo credit: Anissa Reed)

The second fish tested was the jaundiced chinook salmon. (Photo credit: Anissa Reed)

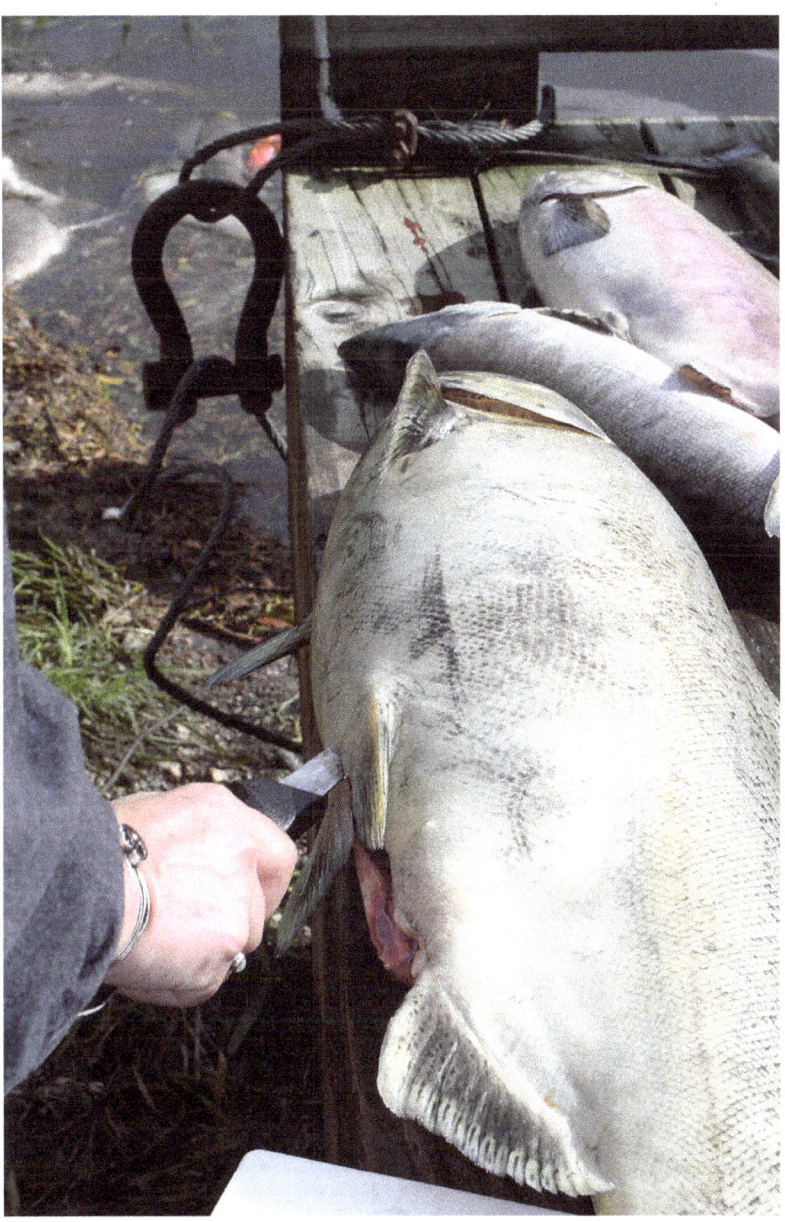

Chinook salmon are the largest of the five Pacific salmon. They weigh on average 30 pounds, although some have grown to over 100 pounds. This is remarkable since this species only lives for an average of five to seven years, although some live for up to eight years. (Photo credit: Anissa Reed)

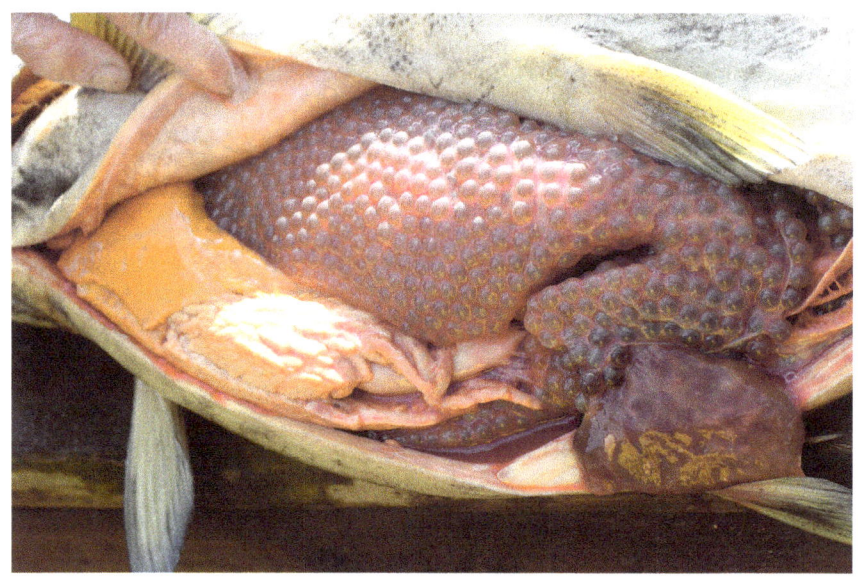

Chinook #2 was full of eggs and its liver was an abnormal yellow.
In chinook, jaundice is a symptom of Piscine orthoreovirus, the
Norwegian Atlantic farm salmon virus. (Photo credit: Anissa Reed)

A close-up of a jaundiced liver, something that is often seen in wild
salmon that have died before spawning. (Photo credit: Anissa Reed)

Acknowledgements

Agreat big thanks to Commissioner Bruce Cohen, his entire staff, and all of the legal counsel and their teams. Even though I was an independent observer of the proceedings, I was made to feel more than welcome, which made my work so much easier and more enjoyable. Special mention goes to Leonard Giles, who served as the registrar and gave me the freedom to adjust the room and lighting so that I could get the best images possible within the Inquiry's terms of reference. Also special mention to Carla Shore, director of Communications for the Inquiry, for her guidance and help throughout my time at the hearings.

I would also like to thank all the participants who feature in the film. It cannot have been easy to stand up and answer so many tough questions. I hope this book and the film will be useful for anyone who finds themselves on the stand at a Canadian inquiry in the future.

I would especially like to thank Alexandra Morton for her support of the film. Her time on the panel was challenging in many ways, and I hope it wasn't too difficult to relive it when the film, *The Unofficial Trial of Alexandra Morton*, came out. I also would like to thank her for contributing to this book and giving the public her *perspective* on the Inquiry itself. My thanks also go to her lawyer, Gregory McDade, who features prominently. My position in the courtroom was quite close to his eye line and it must have been a special challenge for him to focus with a camera pointed right at him. There was no easy way to change the situation since both our positions were pretty much designated and determined by the constraints of the room. I thought he did a masterful job with this extra challenge during an intense couple of days with his client on the panel.

A big thanks to Canada. For all our warts, we still believe in freedom of speech and the rule of law, even when we fight about it. Canadian freedoms allowed an independent film maker to record a very real struggle between stakeholders in one of its federal courtrooms. I hope Canada will continue to show the world that differences can be settled through debate and discussion rather than with violence.

I would like to thank my entire team for their hard work and dedication during the making of the film. It was an enormous task to go through all of the testimony and extract the key parts and distill them into a very watchable film. In particular, I would like to mention Rob Neilson, who pulled together the post-production team and pushed the film over the finish line. This was not an easy film to cut together, as there was only one camera allowed in the courtroom and the many sudden camera moves posed a special challenge in the edit room.

A huge thanks to Jan Westendorp for her amazing skill and guidance in managing the design and publishing of this book. And to Lesley Cameron, for her terrific and prompt story editing, which kept the book project moving forward and a dream project to work on, not to mention the numerous times she rescued my grammar and sentence structure.

And of course, many, many thanks to my family, who put up with a pretty intense year of filming. I was gone early each morning and worked late into the night, backing up each day's footage. It was an enormous task, and I was absent from many family gatherings in 2011. They often mentioned to me that I was losing weight and needed to slow down. It turned out it wasn't the Inquiry that was causing the weight loss, but an undiagnosed bout of Lyme disease. (Stay tuned for more on that at a later date.)

References

Exhibits

The following list contains the exhibit numbers and descriptions of all the exhibits that were presented to the Cohen Inquiry during Alexandra Morton's two days of testimony.

Exhibit #	Description
1540	Dill, L. (2011, June). *Impacts of salmon farms on Fraser River sockeye salmon: Results of the Dill investigation.* Department of Biological Sciences, Simon Fraser University. Cohen Commission Technical Report 5D. Commission of Inquiry into the Decline of Sockeye Salmon in the Fraser River.
1557	Morton, A. Routledge, R., McConnell, A., & Krkosek, M. (2011). Sea lice dispersion and salmon survival in relation to salmon farm activity in the Broughton Archipelago. *ICES Journal of Marine Science, 68*(1): 144–156.
1798	Curriculum vitae of Alexandra Morton.
1799	Curriculum vitae of Catherine Stewart.
1800	Curriculum vitae of Clare Backman.
1801	Curriculum vitae of Mia Parker.
1802	Salmon Aquaculture—Comparison of Regulations.

1803   *Protection, restoration and enhancement of salmon habitat*: Focus Area Report, Norway.

1804   Team of the Office of the Commissioner for Aquaculture Development. (2004). *Recommendations for change: Report of the Commissioner for Aquaculture Development to the Minister of Fisheries and Oceans Canada.* Office of the Commissioner for Aquaculture Development.

1805   Parsons, W. (2011). *Perspective on the technical challenges associated with closed system aquaculture for grow-out of salmon of B.C.* Prepared for the Aquaculture Judicial Inquiry on behalf of the BC Salmon Farmers Association.

1806   Riddell, B. (2006, December 8). Email to Alan Cass et al. Subject: Cultus Lake prespawning mortality 2006. Science Branch, Pacific Biological Station, Nanaimo, BC. Department of Fisheries and Oceans.

1807   Whitehouse, T. (2006, November 10). Email to Keri Benner and Paul Welch. Subject: Cultus diagnostic update.

1808   Willis, D. (2008, August 26). Email to Brian Leaf et al. Subject: Prespawn update: Upper Pitt (as of August 25). Department of Fisheries and Oceans.

1809   Patterson, D. (2007, January 31). Email to Kristi Miller-Saunders. Subject: Cultus Lake sockeye salmon histology samples—gill form of a Parvicapsula-like parasite is present. Department of Fisheries and Oceans.

1810   Pacific Biological Station. (2009, December 21). Memorandum to Rick Stitt and Doug Lofthouse, Weaver Creek Spawning Channel. Subject: PSM loss investigation Oct 29, 2009 Summary Report. Department of Fisheries and Oceans.

1811   Patterson, D., & Bradford, M. (2007). DFO internal memorandum. Late run research on pre-spawning mortality.

1812   Thompson, B. (2009, May 24). Email to M. Higgins,
        C. MacWilliams, D. Lofthouse, W. Bennett, D. Celli, and
        J. Hwang. Subject: Nadina gill samples. Department of
        Fisheries and Oceans.

1813   Miller, K. (2008). *Physiological control of entry timing and
        fate.* Fisheries and Oceans Canada.

1814   Garver, K. (2011). *Diseases and viruses present or investigated
        that might cause mortality in BC Pacific Salmon.* Paper pre-
        pared for Cohen Commission.

1815   Miller-Saunders, K. (2009, October 2). Email to C. Parken.
        Subject: Clarification question, with attachment titled,
        2009 Fraser Sockeye meeting hypothesis. Head, Molecular
        Genetics Section, Pacific Biological Station, Nanaimo, BC.

1816   Miller-Saunders, K. (2009, October 5). Email to Mark
        Saunders. Subject: Briefing report. Head, Molecular
        Genetics Section, Pacific Biological Station, Nanaimo, BC.

1817   Miller-Saunders, K. (2009, November 4). Email to Mark
        Saunders. Subject: Version 2. Head, Molecular Genetics
        Section, Pacific Biological Station, Nanaimo, BC.

1818   Tucker, S., Trudel, M., Welch, D.W., Candy, J.R., Morris,
        J.F.T., Thiess, M.E., Wallace, C., Teel, D.J., Crawford, W.,
        Farley Jr., E.V., & Beacham, T.D. (2011). Seasonal stock-
        specific migrations of juvenile sockeye salmon along the
        west coast of North America: Implications for growth.
        Transactions of the American Fisheries Society, 138(6):
        1458–1480.

1819   Jones, S. (2008, October 8). Email to Kristi Miller-Saunders.
        Subject: Preservation of RNA for functional genomic
        studies: A multidisciplinary tumor bank protocol. Pacific
        Biological Station, Department of Fisheries and Oceans,
        Nanaimo, BC.

1820   Miller-Saunders, K., & Lake, D. (2010, November 16). Email string between Lake and Miller-Saunders re "Media lines—fish disease", Department of Fisheries and Oceans.

1821   Johnson, S. (2009, November 2). Email string between S. Johnson, A. Tompkins, M. Saunders, S. Jones, K. Garver. Subject: "brief summary needed related to Sx response." Department of Fisheries and Oceans.

1822   Constantine, J. (2006, January 3). BC Ministry of Agriculture and Lands Memorandum, Report from Meetings with Mainstream Re: occurrence of Piscirickettisa salmonis at sites in Broughton. Province of British Columbia.

1823   Palermo, V., & Hargreaves, B. (2005, April 27). *Detection and distribution of significant clusters of sea lice* (Lepeophtherius salmonis *and* Caligus *sp.*) *infestation from samples of juvenile salmon and stickleback in the Broughton Archipelago, Knight Inlet, BC. 2003–2006 using a spatial scan statistic* (SaTScanTM) [DRAFT].

1824   Davis, T., Sloan L., et al. (2009, August 19). Email string between them. Subject: "Brian Riddell article." Department of Fisheries and Oceans.

1825   Hargreaves, B. (2009, October 16). Email to M. Saunders. Subject: Favour lice. Attach: SeaLice BCSFA sea lice reports in 3–2 (2004–08) BCMAL compiled October 2009 with Brent's additional graphs.xls; Map—Fish Farms locations and BC MAL fish health sub-zones.doc. Department of Fisheries and Oceans.

1826   Klaver, M. (2009, October 23). Email string between A. Thompson, B. Hargreaves, et al. Subject: RE: Sea lice data request for industry. Department of Fisheries and Oceans.

1827   Gillis, D. (2009, October 26). Email string between
        H. James, M. Saunders, et al. Subject: re Qs and As for
        meeting with the FCC on Oct. 29. Department of Fisheries
        and Oceans.

1828   Beamish, R. (2009, October 26). Email to Mark Saunders.
        Subject: Harrison sockeye. Pacific Biological Station,
        Department of Fisheries and Oceans, Nanaimo, BC.

1829   Hargreaves, B. (2010, January 27). Email to Mark Saunders.
        Subject: Latest version of Fraser sockeye work plan.
        Department of Fisheries and Oceans.

1830   Power, J. (2009, December 1). Email string between
        E. Porter, C. Wong, A. Thompson, et al. Subject: A. Morton's
        comment on egg imports. Aquaculture Management
        Directorate, Fisheries and Oceans Canada.

1831   Shea, G. (n.d.). Draft memorandum to Cabinet, Partnership
        Fund to Pilot Closed Containment Aquaculture
        Technology. Fisheries and Oceans Canada.

1832   Morehouse, B., & McLaren, M. (2009, September 21). *The
        Fisheries Act and fish & ecosystem health management activities.* DRAFT Discussion Paper on the Need for Greater
        Clarity of Governance for Subsection 36(3) and Section 32
        of the *Fisheries Act.* Fisheries and Oceans Canada.

1833   Fisheries and Oceans Canada. (2010). Aquaculture
        Innovation and Market Access Program 2009–10. Funding
        Table by Region.

1834   Fisheries and Oceans Canada. (2011). Aquaculture
        Innovation and Market Access Program 2010–11. Funding
        Table by Region.

1835   Salmon, R. (2010). Office of the Commissioner of Lobbying
        of Canada, lobbying statement for CAIA. Government of
        Canada.

1836   Swerdfager, T. (2010). CAIA/DFO California Trip Report. Department of Fisheries and Oceans, Government of Canada.

1837   Coastal Alliance for Aquaculture Reform (CARR). (2011, May 27). Letter to Mr. Schuessler and Minister Ashfield.

1838   *Better than the rest? A resource guide to farmed salmon certifications.* (April 2011). Coastal Alliance for Aquaculture Reform and Living Oceans Society.

1839   Morton, A. (2011, August 31). Cohen Inquiry Aquaculture Hearings. *Resist Extinction.*

1840   Morton, A. (2011, September 5). Unwanted trespass!!!! *Resist Extinction.*

1841   Boulet, D., Stuthers, A., & Gilbert, É. (2010). *Feasibility study of closed-containment option for the British Columbia aquaculture industry.* Innovation and Sector Strategies Aquaculture Management Directorate, Fisheries and Oceans Canada.

1842   Swan, L. (2011, May 20). Letter to Alexandra Morton Re: Preventing the spread of infectious diseases in aquatic animals. Canadian Food and Inspection Agency, Government of Canada.

1843   Karreman, G. (2010, April 29). Email chain between J. Power, M. Sheppard, S. Ford, et al. Subject: Re: Effluents and new CFIA and/or DFO regulations.

1844   Thompson, A. (2008, June 3). Email to M. Klaver. Subject: Re: Effluent: processors and vessels.

1845   Wright, A., & Arianpoo, N. (2010, May). *Technologies for viable salmon aquaculture: An examination of land-based containment aquaculture. A public domain report* [Draft report]. Submitted to the SOS Solutions Advisory Committee.

Documents that Appear or Are Mentioned in the Film

Canadian Wildlife Federation. (2008, November). Roland Michener Conservation Award to Alexandra Morton for Outstanding Conservation Achievement [Certificate]. Ottawa, Canada.

Chamut, P.S. (1994, February 14). Letter to A. Morton. Re: Outbreaks of furunculosis in salmon cultured in the Broughton Archipelago. Department of Fisheries and Oceans Canada.

Farms threaten key gillnet fishery. (1993, June 21). *Fisherman*.

Griffen, S. (1994, July 25). Alexandra Morton: A watchful eye on salmon farming. *Fisherman*.

Hinkson, Justice. (2009, February 9). British Columbia (Agriculture and Lands) 2009 BCSC 136 between Alexandra B. Morton, Pacific Coast Wild Salmon Society, Wilderness Tourism Association, Southern Area (E) Gillnetters Association, and Fishing Vessel Owners Association of British Columbia and Minister of Agriculture and Lands, The Attorney General of British Columbia and Marine Harvest Canada, Inc.

Krkosek, M., Ford, J.S., Morton, A., Lele, S., Myers, R.A., & Lewis, M.A. (2007). Declining wild salmon populations in relation to parasites from farm salmon. *Science, 318*(5857): 1772–1775.

Morton, A. (2000). Occurrence, photo-identification and prey of Pacific white-sided dolphins (*Lagenorhynchus obliquidens*) in the Broughton Archipelago, Canada, 1984–1998. *Marine Mammal Science, 16*(1): 80–93.

Morton, A. (2011, August 14). *What is happening to the Fraser sockeye?* Aquaculture Coalition [Report based on documents submitted to the Cohen Commission].

Morton, A. (2011, September 8). On the stand at Cohen. *Resist Extinction*. https://alexandramorton.typepad.com/alexandra_morton/2011/09/index.html

Morton, A., & Symonds, H. (2002). Displacement of *Orcinus orca* (L.) by high amplitude sound in British Columbia, Canada. *ICES Journal of Marine Science, 59*(1): 71–80.

Morton, A., & Volpe, J. (2002). A description of escaped farmed Atlantic salmon *salmo salar* captures and their characteristics in one Pacific salmon fishery area in British Columbia, Canada, in 2000. *Alaska Fishery Research Bulletin, 9*(2).

Siddon, T., & Savage, J. (1988, September 6). Canada/British Columbia Memorandum of Understanding on Aquaculture Development. Agreement between The Government of Canada represented by the Minister of Fisheries and Oceans and the Government of British Columbia represented by the Minister of Agriculture and Fisheries.

Simon Fraser University. (2010). Doctor of Science, Honoris Causa to Alexandra Morton [Certificate].

www.ingramcontent.com/pod-product-compliance
Lightning Source LLC
Chambersburg PA
CBHW051513120626
46551CB00012B/896